FOREST RESOURCES CONSERVATION AND REGENERATION

Dr. (Mrs.) Anita Roy Mukherjee is M.A. Ph.D. in Geography, Calcutta University. She was a research fellow of I.C.S.S.R. She worked as a senior research fellow in the Central Board for the Prevention and Control of Water Pollution. Formerly she was the lecturer of Geography in Gokhale Memorial Girls' College, Calcutta. Now the author is engaged as the Head of the Department of Geography in a degree college in Domjur, Howrah. At present she has undertaken a Project sanctioned by the I.C.S.S.R. on "Social forestry an instrument for the developement of rural economy in Midnapur District (W.B.)—A case study". She has to her credit a number of articles in reputed journals of geography.

She has a long experience in rural economics, rural development etc. The study includes door to door survey in the forest villages. This uphill task was completed by the author through prolonged painstaking field works in these remote areas.

FOREST RESOURCES CONSERVATION AND REGENERATION

A STUDY OF WEST BENGAL PLATEAU

ANITA ROY MUKHERJEE
Department of Geography
Azad Hind Fouz Smriti Mahavidyalaya
Domjur, Howrah

CONCEPT PUBLISHING COMPANY, NEW DELHI 110059

The Indian Council of Social Science Research gave the financial support towards the publication of this work. However, the responsibility for the contents of the book facts, opinions, conclusions, etc.—is entirely that of the author and not of the ICSSR

ICSSR Consultant : **R.P. Misra**

ISBN 81-7022-562-0

First Published 1995

Published and Printed by
Ashok Kumar Mittal
Concept Publishing Company
A/15-16, Commercial Block, Mohan Garden
NEW DELHI-110059 (India)

TO

MY BELOVED MOTHER

ACKNOWLEDGEMENTS

Words are not sufficient to express my deep sense of gratitude to my teacher, and mentor Dr. Bireswar Banerjee, Prof. of Geography, Calcutta University. His keen interest in my work was enough to sustain my work. I have to mention my deep respect to Dr. (Mrs) Jayati Hazra, Lecturer, Department of Geography, Calcutta University. Without whose kind assistance the work would not have taken to present shape.

I am immensely grateful to Dr. Ajit Banerjee, Forest Expert, World Bank, Washington. He had personally persuaded me to undertake this investigation.

I wish to express my deep gratitude to Mr. Udayan Banerjee, Chief conservator of Forests (Social Forestry), Government of West Bangal for his valuable suggestion, constant inspiration and encourgement.

I am deeply indebted to Mr. Somenath Mukherjee, Assistant Forest Officer, Monitory and Evaluation Cell, for making available to me the various facilities in forest areas during my field survey.

I wish to express my appreciation to Mr. Pranabesh Sanyal, D.F.O. Burdwan for the facilities made available by him.

I wish to express my gratitude to Mr. Shyamal Burman, Forest Economist, Mr. Kalyan Chokravarty D.F.O., Planning and Statistical Cell, Mr. Sanat Das, D.F.O., Purulia, Directorate of Forests, Government of West Bengal, for their help and assistance.

I am also indebted to Mr. S.C. Dey, Joint Director, Forest Survey of India, Government of India.

The help rendered by Mr. Gopal Dey, Statistical Officer and Mr. Bansi Manna, Public Relation Officer, Directorate of Forests, Government of West Bengal in this work are gratefully acknowledged.

Of my friends, I specially acknowledge Dr. Malay Mukhopadhyay and Dr. (Mrs.) Sutapa Mukhopadhyay of Biswabharati University, Shantiniketan.

ANITA ROY MUKHERJEE

CONTENTS

LIST OF FIGURES

LIST OF TABLES

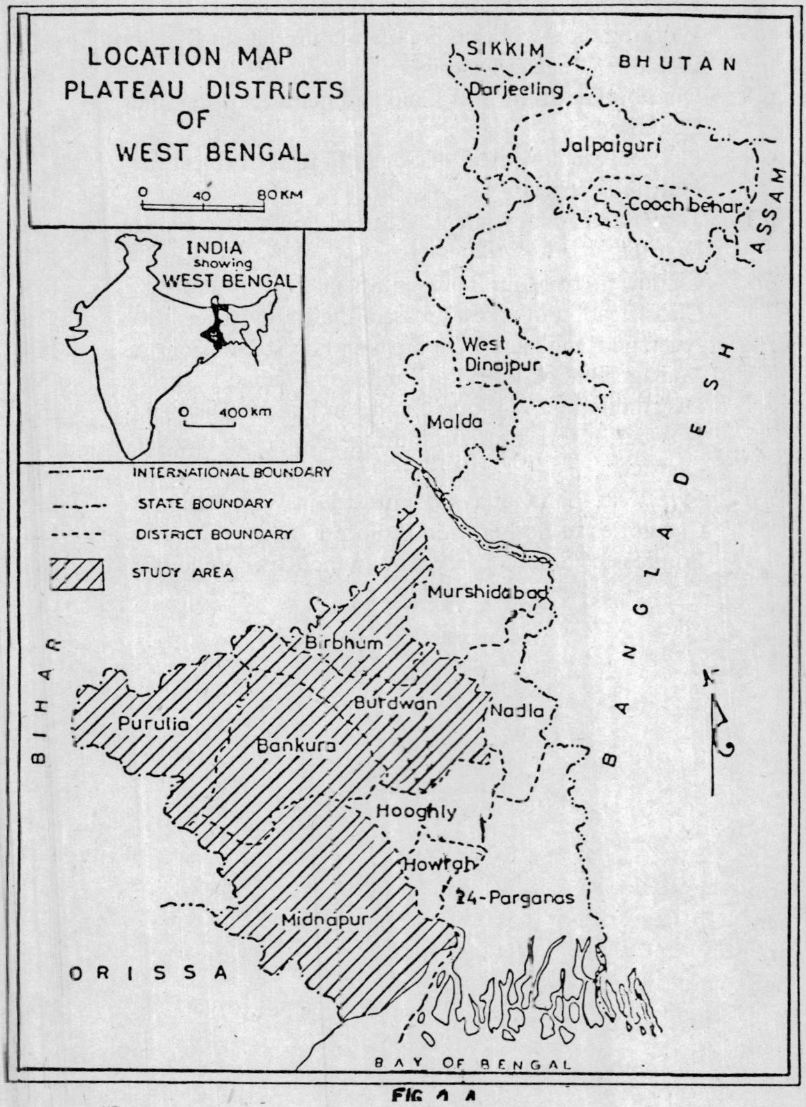

Fig. 1.1: Location Map Plateau Districts of West Bengal

1

INTRODUCTION

General Outline

The plateau districts of West Bengal is located in the western part of the State, comprising of Purulia district and western portions of Midnapur, Bankura, Burdwan and Birbhum districts. The area is bounded on the west by the States of Bihar and Orissa, on the east by the districts of Howrah, 24-Parganas, Murshidabad, and on the south by the Bay of Bengal (Fig. 1.1)

It is interesting to note that extensive portion of the plateau has been buried under the thick alluvium, particularly the eastern and southern parts of the region. The western part of the plateau districts is the extension of Chota Nagpur Plateau.

West Bengal is not particularly rich in forest wealth. On the contrary, it is rapidly facing depletion. This is because of rapid growth of urban areas, high density of population and extension of farming areas in recent years. Covering only 2.7 per cent of the total land mass of India, West Bengal presently supports only 1.6 per cent forest area of the nation. The proportion of forest area in West Bengal is also less than that of India. While 22 per cent of the land area of India is under forest, the figure for West Bengal is only 13 per cent. On the basis of satellite photographs i.e., imagery, the forest area of West Bengal has been estimated to be even less. It is only 10 per cent of her total area.

The per capita forest area of India is 0.14 ha, the corresponding figure for West Bengal is only 0.03 ha. It is, therefore, evident that the forest cover of the State is much less. It should be around one-third, which is considered the minimum necessity to maintain ecological balance, as per the findings of the National Commission on Agriculture.

The forest areas of West Bengal are not uniformly distributed. On the other hand, they show marked regional concentrations. The forests here are particularly concentrated in three distinct regions, namely (i) the hills and foot-hills of Darjeeling Himalaya, (ii) the mangrove forests of Sundarban in the deltaic tracts of the South East and (iii) the hills and plateau of the west. Of these regions the plateau section claims 38 per cent, the mangrove forest claims 37 per cent and Darjeeling hills and foothills claim 18 per cent of the total forest area of the State and 7 per cent are in remaining districts. But in respect of the proportion of foliage cover the forests of the Darjeeling hills and foothills cover nearly one-fourth of the total area of Darjeeling district. The corresponding figure for the plateau districts are only 11 per cent.

Of the three regions of concentration mentioned above, the Plateau districts (Fig. 1.2) have suffered extensively during the past few decades. Growth of mining, industries and new settlements have led to continuous encroachment on the forest areas leading to gradual depletion of the forest wealth. Moreover, the region has a high proportion of tribal population whose economy and ways of life are intimately associated with the forests. These are backward areas where most of the people live below poverty line and quite naturally they depend on forests not only to meet the needs of the fuelwood but also for other necessities of life. It is interesting to note that degeneration takes place quickly in this region. Natural process of regeneration takes time due to environmental constraints arising from degraded soil, variable rainfall, extremes of temperature, fluctuating groundwater level etc. As a result the quality of timber available in the forest is gradually degenerating. Much of the virgin forest areas have been disturbed being replaced by coppice vegetation. The forests of Darjeeling hills and foothills and Sundarban, on the other hand, are in a better position because of congenial climate and terrain and comparatively lesser human interference than the plateau districts. As such, the natural process of regeneration is high in the former regions. The forests of the plateau districts, therefore, requires better management techniques both in respect of conservation and regeneration of this valuable wealth.

From environmental point of view, the forests of the plateau districts play more significant role than in other areas of the States. It is well known that this region suffers from soil erosion which is a

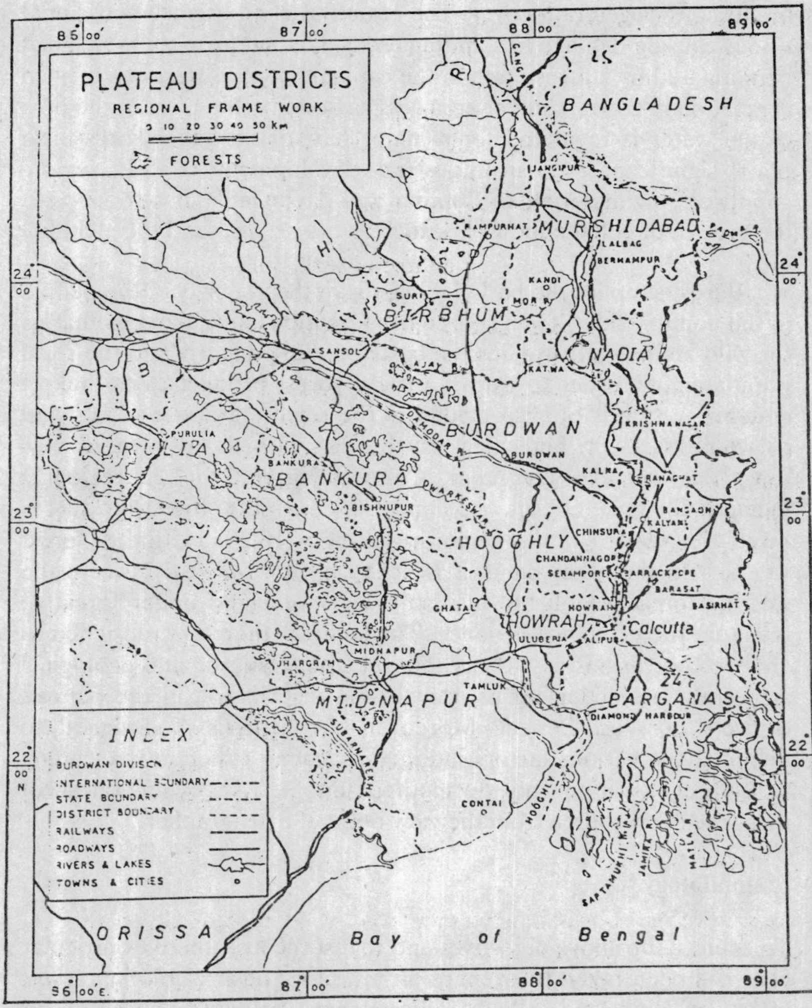

Fig. 1.2: Plateau Districts: Regional Framework

natural hazard of great magnitude but still it can be considerable controlled by the systematic growth of vegetation and preservation of the existing vegetation cover. The region also experience occasional floods causing damage to standing crops, life and property which can be mitigated by afforestation in the catchment areas. Growth of such vegetation in the catchment area acts as flood cushion where the flow of the water is regulated for a major part of the year. Forest also plays significant role in industrial development of this region, supplying raw materials to Lumber and paper mills. It also helps in the growth of cottage industries such as tassar, lac, seed oil, alchohol etc.

Forest supplies the basic fuel needs of the majority of the people in the country side. It is a source of recreation as well as sanctuaries for wild animals. It supplies the basic building materials to the rural population. In order to extract various forest products, vital link of communication is often provided in the remote areas. The demand of wood is infect increasing day by day with the growth of new townships, such as Durgapur, Chittaranjan etc. So the region is facing a dilemma. On the one hand, the demand of forest product is rapidly increasing in this region resulting in fast depletion of forest wealth. On the other hand, it is necessary not only to preserve the existing forest wealth but also to bring more land under forest to maintain the ecological balance. The study, is therefore, designed to analyse the various aspects of this valuable resource and ecological attributes interacting on its growth and distribution in the plateau districts of West Bengal. Moreover, the analysis is designed to decipher the various factors interacting amongst themselves leading to the degeneration and the methods adopted for regeneration of this valuable resource from the view point of a geographer.

Methodology

To achieve the above objectives and to test the hypotheses, particular care has been taken to prepare an inventory of available materials, maps on this and allied topics from various published and unpublished as well as official sources. Macro-level data have been collected from the Directorate of Forests of the State Government, whereas micro-level data have been compiled from various Divisional Forest Offices, Ranges and Beats. The work is also based on information gathered from extensive and intensive field-work both in accessible and remote areas. In-depth analysis of the role of natural environment on the distribution of forest areas has been carried out. Regional variations of forest types and their problem have also been identified.

The terrain pattern of the region has been studied from 1:50,000 topographical sheets of the Survey of India. The climatic data for 30-year period have been collected from the Regional Meteorological Office Government of India. The groundwater data have been collected from the records of the Central Groundwater Board, Government of India. The data on soils have been collected from Directorate of Soils and Landuse Survey and also from Directorate of Agriculture, Government of West Bengal. The spatial distribution of forests have been identified and mapped from landsat imagery.

The data thus collected from various sources have been systematically processed, classified and computed. Some of them are used in drawing representative maps and cartrograms.

Intensive field studies have been carried out in five villages located in Midnapur and Purulia districts, in order to assess the influence of forest on the village economy.

This analysis is, therefore, based on both empirical as well as on quantitative techniques which are supplemented by frequent field checks.

Previous Works

Works on the forests of this State have been carried out by persons directly attached to the forest department. Naturally most of these works are botanical in nature. The 'Report of the Forest Committee, (1939)' from the Department of Forest and Excise is the first published report describing the status of the forest wealth of the State. The *Indian Forest Ecology* of Puri (1960) is a more systematic study of forest, indicating the role of ecological factors on its growth and types, with specific example from Central India. Champion and Seth (1968) have made a systematic classification of forests which does not throw much light on the role of environmental parameters on such classification. The Centenary Commemoration Volume (1964) of the Forests of West Bengal is an excellent assemblage of papers which depict new information on the status of forests in the plateau section. Das (1954) has made a very valuable contribution on the forests of Assam is a pioneer work on the topics done by a geographer. But the work pertains to Assam having little relevance to the present study area.

Papers presented at several silvicultural conferences have been published, some of them carry valuable information on the forests of the plateau section relating to the special fields of the researchers. Valuable works have also been published in *Indian Forester*, but these are not sufficient enough to make a realistic assessment about forests of this area.

It has, therefore, been apparent that most of these studies have been carried on by non-geographers. These are mainly related to taxonomy, silvicultural practices etc. where the role of natural and social environments on forest growth and its spatial distribution, its present problems etc. are largely lacking.

Some researchers have recently devoted considerable time in drawing vegetation maps of various regions of India (Gaussen, 1959). Some others have carried work on the role of various environmental parametres on forest growth in selected regions of the country (Meher Homji, 1980). The National Atlas and Thematic Mapping Organisation has brought out a well documented *Forest Atlas of India* showing the spatial distribution of forest in various regions of the country. But the maps are based on back dated data. Mukherjee (1983) has carried investigation on the impact of social forestry on community development in selected parts of the region.

It is, therefore, apparent that the researcher has to carryout work within the constraint of scanty publications on this or allied topics.

Natural Environment

Terrain Pattern

Major part of the study area consist of flat or gently undulating monotonous alluvial plain, with elevation less than 300 m. The plain rises gradually towards the west, while towards the south, the land surface decends to the Bay of Bangal through Midnapur coastal plain. The continuity of the landform is disturbed by number of west to east flowing rivers (Fig. 1.3).

The 75-metre contour line roughly divides the region into two longitudinal divisions with distinct lithological features. The plains of eastern Birbhum, Burdwan and Midnapur are composed of

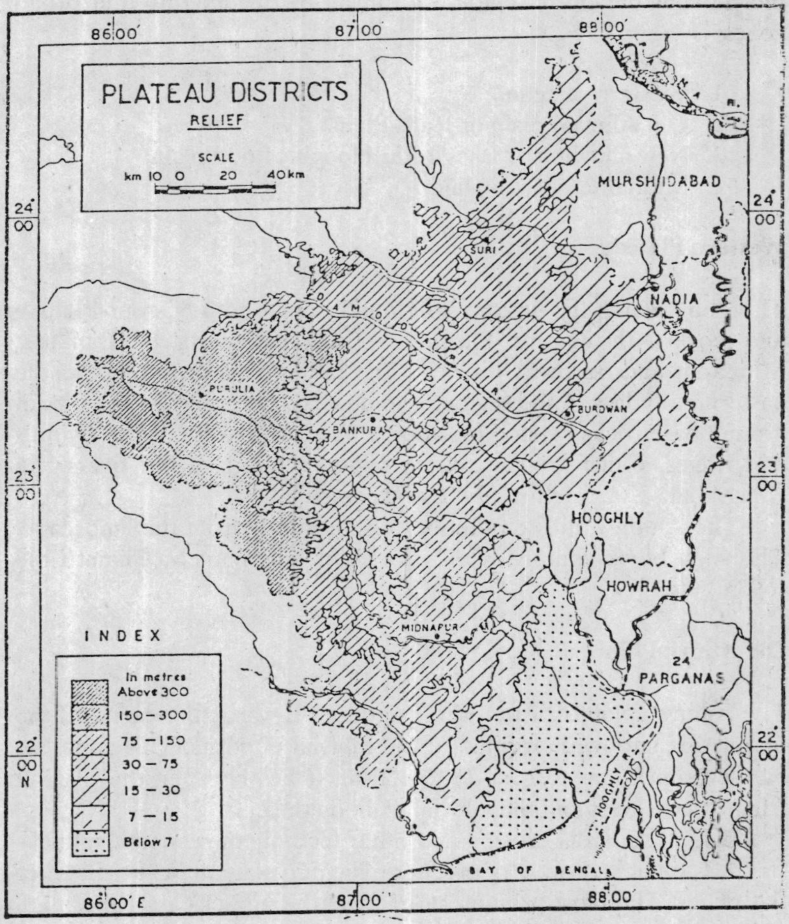

Fig. 1.3: Plateau Districts: Relief

transported alluvial deposits, usually lying below 75 m elevation, while the uplands in the western parts of this area are composed of hard rocks, laterites and lateritic soils formed *in situ*.

Depending on uniformity in structure, relief, lithology and morphology the entire region can further be divided into four broad zones:

1. Western plateau.
2. Old Alluvial plain or Rarh Plain.
3. New alluvium or Bhagirathi-Hooghly flood plain.
4. Midnapur coastal plain.

Western Plateau

This zone marks the eastern continuation of Chota Nagpur Plateau and covers the entire Purulia district and parts of Bankura, Midnapur and Burdwan districts. The 150 m contour line marks the eastern boundary of the region. The underlying formation consists of granite base rocks capped by laterite. The plateau is marked by hills and ridges, which are the source region is some of the rivers like Kangsabati, Kumari, Silai and Darakeswar.

The Ajodhya hills on the west rise abruptly from the landscape. This high block comprises of a series of hills namely, Chamtu hill (698m) etc.

Old Alluvial Plain or Rarh plain

The rolling uplands of the west continues towards the east and the undulating surface is formed of old alluvial deposits. This zone has an elevation ranging from 75 to 150 m. It lies between the plateau fringe on the west and the alluvial plain on the east.

Due to subareal erosion and other tectonic movements, hillocks stand as detached outliers of Chota Nagpur plateau in the midst of flat plains. The most prominent amongst the hillocks are Biharinath (440m) and Susunia hills (440 m).

The former contains sedimentary rocks of Gondwana system, while the latter is mainly of gneissic and schistose rocks of Archean age, except the top of the hills which is formed of felspathic quartzite.

A small segment of the Rarh is also found on the west of

Subarnarekha river in Midnapur district with older deposition and a continuation of the dissected Chota Nagpur plateau.

New Alluvium or the Bhagirathi-Hooghly Flood Plain

The New Alluvial plain is a typical depositional plain stretches from Birbhum in the north up to the Bay of Bengal and Orissa on the south. The average elevation in 75m on the west and 7m on the east. Towards the south this plain gradually merges with the Damodar Rupnarayan Delta. In this section rivers have tidal influences. On the basis of tidal influence the region is further sub-divided into following three sections :

1. Plains of north or Bhagirathi-Hooghly Plain.
2. Hooghly Damodar Rupnarayan Delta.
3. Midnapur Plain.

TABLE 1.1 : Correlation between Relief and Concentration of Forest in the Plateau Districts

Physiographic Regions	Micro Regions	General elevation (in m)	Percentage of area to total area	Percentage of forest to total forest area	Percentage of forest in each micro region
Physio-graphic Plateau	Hills	300	1.44	8.80	74.81
	Uplands	150 – 300	13.12	21.14	15.60
Old alluvial Plain or Rarh plain	Plain	74 – 150	18.63	28.40	14.79
Plains	Bhagirathi Hooghly Plain	30 – 75	28.23	41.66	14.33
	Hooghly Damodar Rupnarayan Delta	15 – 30	12.99	–	–
	Midnapur Plain	7 – 15	9.41	–	–
Midnapur coastal plain	Coastal Plain	7	16.18		
			100.00	100.00	

Midnapur Coastal Plain

The coastal plain of Midnapur stretches between the mouths of the Subarnarekha on the west to the Hooghly river in the east.

Terain Pattern and Distribution of Forests

A large portion of this region is covered with forest with intervening farm lands, wastelands and villages. The region is noted for diversities in topography, geological formations, parent rocks, etc. As a result, the distribution pattern of forests also records variations. For example, while hills and plateaus are substantially covered with forest, the deltas and flood plains of the east and south are devoid of any forest cover which shows the significant relationship between forest and altitude. Moreover, in the new alluvium tracts of the deltas and coastal plains, the original forest cover has already been replaced by systematic cultivation. In the old alluvium of Rarh plain, forests are of scattered location. Most of the forest areas are clothed by Sal as the principal species with sprinklings *Peasal, Simul, Mahua* and *Kend* as its associate. In the western hill and plateau, large tracts of forest areas are common.

Geological Formations

Depending of geology and rock structure, the plateau section can be broadly divided into two divisions :

(i) western part of the area being formed of granite and gneiss, and
(ii) the east which is almost wholly concealed beneath a blanket of alluvium. The Archean terrain of the west is the eastern extension of Chota Nagpur plateau. It is most dominant and oldest rock formation of the plateau districts. The older group of rocks occurring in the west are quartzite, phylite, micaschists etc. A number of dolerite and metadolerite dykes occur as intrusives in Purulia. Otherwise the plateau is covered by granite gneisses. At places these rocks occur as intrusives into the meta-sediments.

Along the Damodar valley particularly in between the Damodar and the Ajay river, occur thick beds of Gondwanas, being represented by the Barakar, Raniganj, Panchet and Supra-Panchet rocks. In between, there are alternate layers of sandstone, shale and coal. The alluvium on the east also differs in age and character from east to west. Towards the west, the old alluvium is mostly transformed into laterites being formed of hydrated oxides of aluminium and iron with small amounts of manganese and titanium. Towards the east and south, the riverine sand, silt and clay form newer alluvium of comparatively recent age (Fig. 2.3).

The geological formations of an area not only influence its physiographic character, but also influence the soil types which sustain the natural vegetation. It has been noticed by the naturalists that there are some relationships between surface flora and subsurface lithology. It is an established fact that Pial (*Buchanania lazen*) is most frequent species on metadolerites, because it is a clay loving species. Similarly Kend (*Diospyros melanoxylon*) is most frequent on mica schists as kend requires coarse sandy soil.

It has been found from an experiment (Sanyal, 1977) that the percentage of miscellaneous species are higher on metadolerites than quartzite and mica schists. The density of sal stems are same on mica schists and quartzite formation and less on the metadolerites.

Drainage Pattern

The drainage pattern of this area depends largely on the morphological conditions, groundwater level, periodicity of rainfall and soil characteristics. The plateau fringe is formed of a number of east and south-easterly flowing rivers, being separated from each other by deep valleys (Fig. 1.4). Most of these rivers have attained maturity or old age.

The most important rivers entering into the plains from the western plateau are the Brahmani, Dwarka, Mayurakshi, Ajay, Damodar, Dwarakeswar, Silai, Rupnarayan, Kasai, Kaliaghai, Haldi, Rasulpur and the Subarnarekha river. The total catchment area of the river covers an area of about 64,000 sq. km. These rivers have their regimes much older than the Ganga. They have formed the old deltaic plain with sand and silt lying to the west of the New Deltaic plain.

The rivers become torrential during the rains. During November to May the flow may drop down to mere trickles. In some cases, the

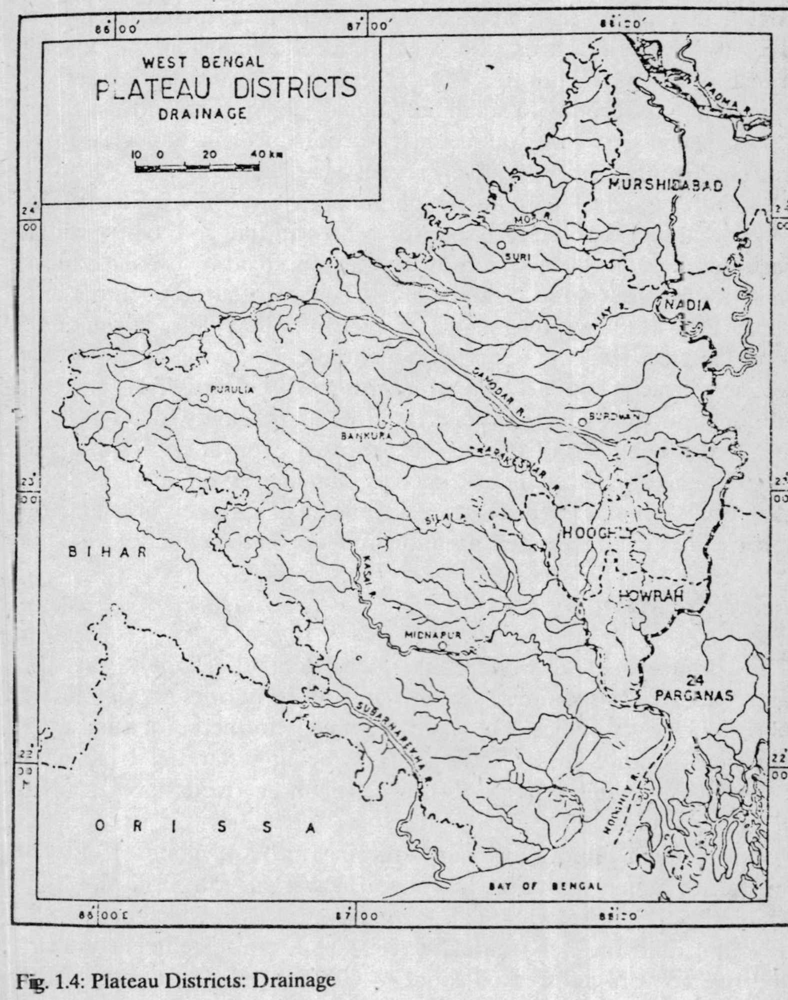

Fig. 1.4: Plateau Districts: Drainage

rivers become practically dry. Over 90 per cent of the total discharge of these rivers take place during June to October, but the peak discharge usually takes place in August, September.

TABLE 1.2: An Inventory of the River Systems of the Plateau Districts

Name of rivers	Length (in km)	Source	Drained into	Catchement area (sq. km)
Mayurakshi	53 in Birbhum	Santhal Pargana	Bhagirathi	5,540
Ajay	138	Bihar	Bhagirathi	2,260
Damodar	595	Chota Nagpur	Hooghly river	5,591
Dwarakeswar and Rupnarayan	42	Purulia	Hooghly river	10,273
Kangsabati and Haldi	442	Purulia	Hooghly river	9,658
Subarnarekha	203	Chota Nagpur	Bay of Bengal	3,102

Drainage Density

The drainage density vary from 0.10 to 0.22 km/sq. km. It is high on the east because the river throws off distributaries in their lower course which does not mean that it is well drained. Low drainage density occurs in the west. Though it is fact that inadequate drainage channels restrict forest growth but the forests of Bagmundi in Purulia is supported by very low rainfall. So no relationship with forest and drainage density can be established.

Relation of Forest with Drainage

Forests usually regulate the drainage of a place. So forest is often called the guardian of the drainage.

In this region the rivers have their sources mainly on the hills and plateaus. So when the hill slopes are kept bare, soil erosion usually gathers momentum, which is a deterrent factor for further forest growth on the other hand this also leads to further accumulation of silts on the other hand this also leads to further accumulation of silts on the river beds and forces the rivers to change their original courses.

In conclusion it can be said that the works of land, river, forest and man are closely inter-linked. Though sometimes they work independently, quite often they have reciprocal influence and form natural equilibrium between man and environment. In plateau districts this interaction leads to disaster, the result of which can be easily visualized. The Bhagirathi Hooghly river in this region is faced with the problem of rapid siltation at its mouth. About a quarter of the areas of the Damodar valley was formerly covered under forest which used to protect the rugged terrain from lashing monsoon rain thereby checking soil erosion. But in recent decades the forests have been extensively removed to make room for industrial use. The case is same with the Mayurakshi, Kangsabati and the Ajay catchments.

Climatic Characteristics

The orography and location of the region results in minor variation in climate of the region. The climatic characteristics of the plateau districts are generally controlled by the monsoons. Bulk of the rainfall is caused by monsoon winds though the amount and distribution are irregular (Fig. 1.5). High temperature with high relative humidity are typical climatic characteristic of the region in summer.

The region experiences three seasons in a year :

1. Hot dry season – March to early June
2. Hot wet season – Mid-June to September
3. Cold Season – October to February.

TABLE 1.3: Relation between Rainfall and Distribution of Forests in the Plateau Districts (1984)

Rainfall (in mm)	Total area (in sq. km)	% age to total area	Forest area (in sq. km)	% age to total forest area	% age of forest in each micro region
1,700	555	1.29	–	–	
1,600 – 1,700	2,095	4.86	–	–	
1,500 – 1,600	3,624	8.42	530	12.68	14.62
1,400 – 1,500	16,179	37.58	1,185	28.36	7.32
1,300 – 1,400	16,155	37.52	1,731	41.42	10.71
1,300	4,450	10.33	733	17.54	16.47
	43055	100.00	4179	100.00	

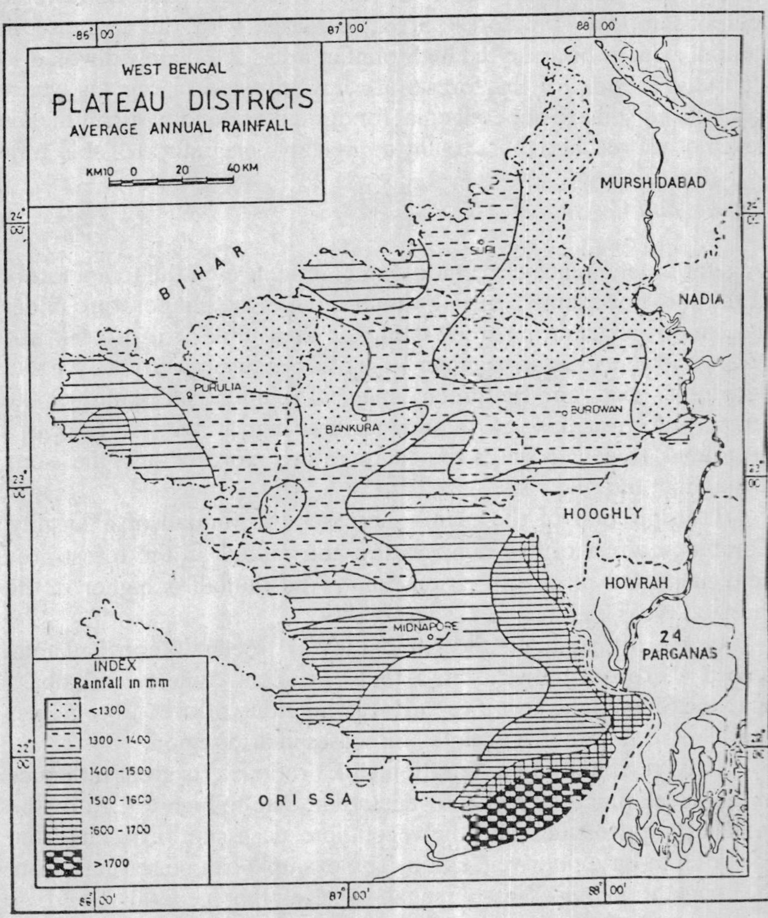

Fig. 1.5: Plateau Districts – Average Annual Rainfall

To analyse the relationship between rainfall and distribution of forests, Pearson's product moment correlation and co-efficient has been work out. The result is -0.3 which indicate its inverse relationship, i.e., the forest area decreases with the increase in rainfall. This is because the high rainfall areas are mainly devoted to agriculture. Most of the forests are now situated in regions where rainfall and soil are not congenial for agriculture and biotic influence have not yet reached there to any appreciable magnitude (Table 1.3).

Climate and Vegetation

Vegetation of a place exert moderating influence on the temperature of the soil. For example during winter months the temperature under the canopy of casurina trees is higher than under sal or teak, the reverse is the case in summer. This is because casurina is not a deciduous but an evergreen species. Under sal canopy, the temperature remains low in summer and high in winter. Teak is a deciduous tree and keeps the temperature more or less the same throughout the year.

Transpiration of the plants increases the rainfall of a locality. Combined affect of low temperature and rainfall in the forest area help the air to reach saturation point. The rainfall is higher in the forested areas than in open space.

In hot dry summer, the comparatively treeless Ranchi plateau record a drop of humidity to 5 to 10 per cent. Whereas Chaibasa near the Singhbhum forest records the humidity of over 50 to 58 per cent. So the variability of rainfall increases in arid region.

The Precipitation Evaporation (P/E) ratio exerts great influence in the development of soil and vegetation. Due to high P/E ratio, the soils get impoverished and the vegetation succession proceeds from base rich to base poor vegetation. The example can be set here from the vegetation in West Bengal, where the general trend is replacement of kend in favour of sal forest. Though the chief cause of succession in a habitat is climatic but sometimes biotic factors are also responsible for it.

Soil

The soil rank next to climate amongst the environmental factors influencing the vegetation. It provides the basic mosaic on which plants grow. "The nature of the soil and consequently the nature of the rocks, by the disintegration of which the soil has been formed is the most important factor in forestry" (Brandis, 1921).

The vegetation type is highly influenced by the soil character. For example the acidic non-calcareous soil is suitable for sal. The coastal soils on the other hand, are endowed with salt deposition, low water, poor organic matter and nitrogen which are suitable for casuarina.

The soils of western part of the plateau districts are porous and well drained because of the absence of clayey soil in the region (Fig. 1.6). This helps in the percolation of water through the soil easily. Thus with the combination of free drainage, water retaining capacity is low. Alongwith it, in dry periods, the groundwater table lies at a depth of 10-15m from the surface. These factors are not particularly favourable for plant growth. Hence the forest cover is thin in this region.

Table 1.4 depicts the picture that there is a close relation between the distribution of soil type and forest cover. The forests are mostly concentrated in two soil zones laterite and red and brown soil.

TABLE 1.4: Relation of Soils with the Distribution of Forests Plateau Districts

Soil types	Total area (in sq. km)	% age of total area covered by each soil group	% age of forest to total forest area	% age of forest cover in each soil group
Red and Brown	8,320	19	39.20	19.69
Laterite	10,100	23	50.68	20.96
Red loam	4,780	11	7.39	6.46
Sandy loam	17,266	40	2.73	0.66
Clay soil	2,095	5	–	–
Saline soil	309	1	–	–
Coastal loam	185	1	–	–
	43055	100	100.00	

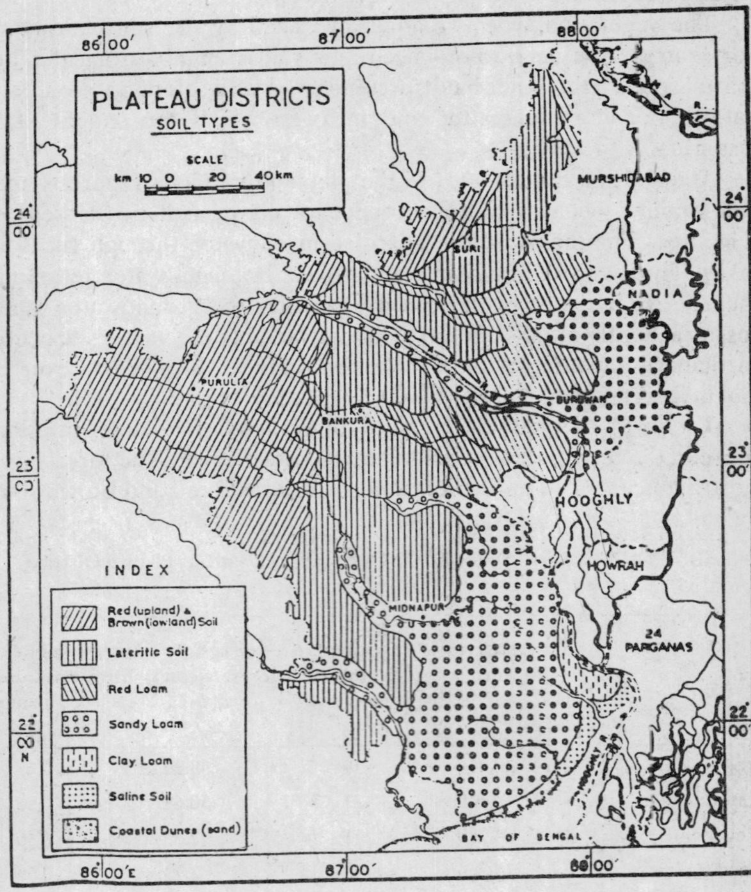

Fig. 1.6: Platau Districts: Soil Types

Soil Moisture

Soil moisture has close relationship with forest. The conservation of soil moisture is needed to boost up plant growth. One of the fundamental effects of vegetation is that it reduces the soil temperature by absorbing a great portion of solar energy through the leaf surface and prevent evaporation from the soil. In an experiment (Ghosh, 1977) carried on in three plots to study the effects of vegetation on moisture regime, it has been found that (a) Bushy Sal Coppice forests retain maximum percentage of moisture (b) Barren wasteland retains the minimum percentage of moisture, and (c) the standing Sal coppice forest stands inbetween the two. The Bushy Sal forest hinders evaporation from soil and encourage transpiration. In case of barren land, there is nothing to check evaporation from the soil. So the moisture content is minimum. The standing sal coppice forest allows the soil to evaporate because of absence of under growth.

Relation of Species with Soil

The soils are the indicators of vegetation types. Different soils record different pH value. The main species grown here is sal. In general the sal forest requires acidic, noncalcareous soil (Yadav, 1966) with varied texture and depth. It grows well where pH value ranges between 5.2 – 6.4. According to Yadav the fine and coarse textured or excessively boulderly soils deficient in moisture retention, are not usually suitable for sal vegetation. The clayey and compact soils having por permeability are not favourable for the regeneration of sal. The mortality of sal is primarily due to moisture deficiency in porous sandy soils. This is particular in regions of low rainfall in Birbhum and parts of Bankura district (Seth, Khan and Yadav, 1960). The teak (*Tectone grandia*) requires calcareous or base rich soils for plantation and natural regeneration. It prefers high pH value, (Puri, 1960).

In case for planted species *Eucalyptus hybrid* of Mysore origin (Mysore gum) is capable of growing on wide range of soil types. The soils suitable for it are generally acidic, non-calcareous with variable texture, organic matter and nitrogen.

Ground water

The surface water usually percolates through joints or cracks and is preserved above the impervious layers of rock. The water preserved in this zone is known as goundwater and the watertable denotes the upper surface of this water. The availability of groundwater depends on the geological character and topography. The rock through which the water moves and the place where it is deposited are very much influenced by the nature and slope of the area. Therefore, the contour of the land determines the point at which the watertable may intersect the ground. Coulson (1940) stated that it depends on widely different factors as topography, geology and climate.

The groundwater-table is important for plant growth only if it is relatively shallow (usually 5 to 10 m), because the major portion of the plant roots are concentrated within this depth. Shallow-watertable is needed for the growth of saplings. The water-table regulates the moisture regime of an area and helps in vegetation growth. Even in areas with low rainfall the availability of moisture to the plants is sufficient if the watertable is within few metres of the surface. On the contrary, the areas may exhibit moisture deficiency if the watertable lies at depth.

Groundwater Condition of Plateau Districts

The Plateau districts have a gradual slope from the western table land to the eastern alluvial plains. So the depth of water varies from the west to east depending on the geological rock structure and soil character (Fig 1.7). Fluctuation of watertable is also noted from the pre-monsoon to post-monsoon months. Thus due to structural and seasonal variations, there occurs a marked change in water-level in the region. The average annual depth of water-level ranges from less than 2 m to more than 6 m. But in summer, the ground-water-level lies from less than 5 m to more than 20 m. The watertable lies close to the surface in the east because of high rainfall and presence of a large number of rivers. At places, the depth is less than 2 m in the coastal areas of Midnapur. Patches of areas in the south can be marked where the depth is less because of the river channels. The aquifer within the depth of 8 to 100m contains saline water, but aquifers above and below this range contain fresh water (Roy. 1973).

Adequate knowledge of groundwater is necessary for the success

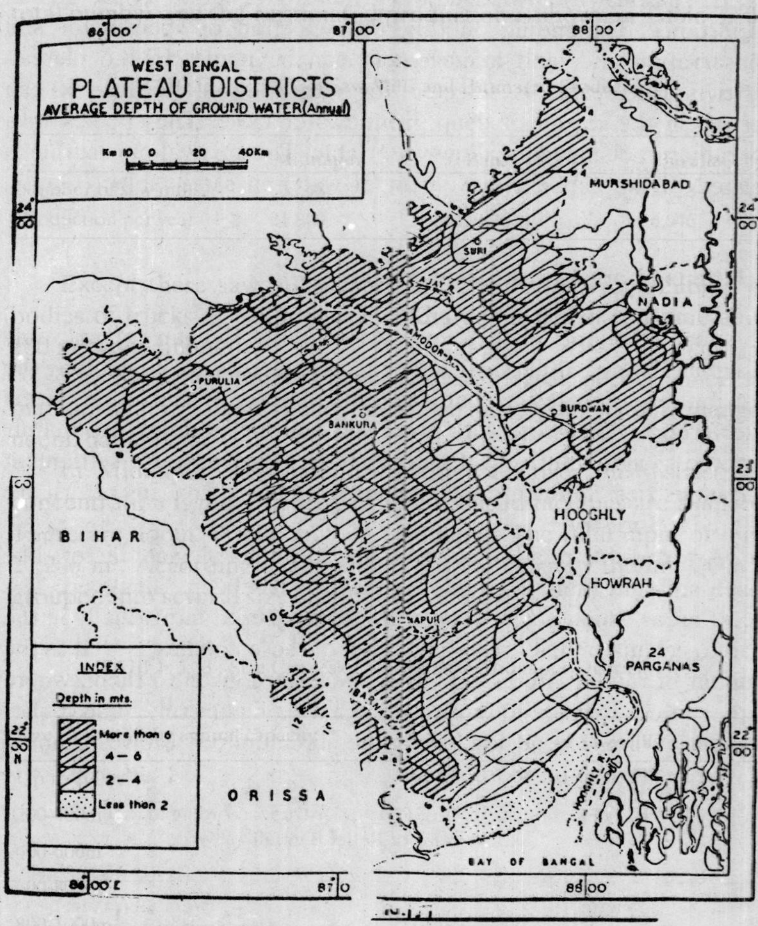

Fig. 1.7: Plateau Districts – Average Depth of Ground Water (Annual)

of any afforestation programme. The requirements of water vary from species to species. In monsoon climate where the rainfall is very much seasonal, knowledge of groundwater-table is of prime importance. Depending on this, the selection of species for soil conservation and land management can be made. In the plateau districts Eucalyptus is selected for afforestation, because it can survive in the region of deep groundwater table. The species with spreading roots needs shallow-watertable. The survival of casurina is related to the capilary movement of water in sandy soils, (Gupta, 1977).

Relation of Forests with Environment

The forests of the plateau sections of West Bengal are unevenly distributed. Several physical and economic factors are controlling the distribution of forestry, but the most important factor is biotic interference. The areas with good soil and high rainfall are mainly devoted to agriculture. Next important areas are used for settlement and industries. So the areas which are not able to give any return are offered to forestry.

Any possible relation between forests and ecological factors like relief, soil and rainfall, chance probability has been calculated. The forest areas which occur in patches are measured individually in the map to calculate their areas. The relief of these patches as well as the amount of rainfall are noted. Score values are given to them, where higher values to given to those forests having higher relief and higher amount of rain. The region having the minimum relief/lowest rainfall is awarded minimum score values.

TABLE 1.5 : Forest Areas of Plateau Districts according to their Score Value of Relief Rainfall

Relief (in metres)	Rainfall (in mm)
300 – 7	1800 – 5
150 – 300 – 6	1600 – 1800 – 4
75 – 150 – 5	1400 – 1600 – 3
30 – 75 – 4	1200 – 1400 – 2
15 – 30 – 3	1200 – 1
7 – 15 – 2	
7 – 1	

The sum total of the minimum score values recorded here is 2 and the maximum possible added score is 12. But the score varies from 4-9 only in this case.

The frequency of forest areas in each group have been computed for relief, rainfall and soil, then the chance probability is calculated. This is indicated by χ^2 and noted the significance level for the different heads. Table 1.6 gives the result of χ^2 test.

TABLE 1.6 : Determination of Chance Probability Test to find Relationship between Ecological Factors and Forest Growth in Plateau Districts

Ecological factors	χ^2	Level of significance
Rainfall	87.85	–
Relief	8.13	0.001
Soil	7.06	0.05

The level of significance is the highest for soil, being followed by relief. However, there is no significant relation between forests and rainfall.

As the level of significance is highest in soil, we have taken the soil types individually. The forest areas under different soils have been grouped where scores are found by adding rainfall and relief score values.

The frequency of scores in different soil groups have then been computed against the area represented by forests in each such group (Table 1.7).

χ^2 test has then been applied to test the level of significance to the Table 1.7 and results have been computed in Table 1.8.

TABLE 1.7 : Frequency of Scores in Different Soil Groups of Plateau Districts

Score	Frequency	Total area	% age of forest area
Laterite Soil			
4	–	–	–
5	1	77.39	3.8
6	5	187.02	9.3
7	21	1085.46	54.6
8	13	657.81	32.7
9	–	–	–
Red and Brown Soil			
4	2	352.20	17.26
5	3	92.79	4.5
6	8	857.74	41.95
7	7	651.37	31.86
8	2	64.50	3.15
9	1	25.80	1.28
Sandy loam soil			
4	–	–	–
5	–	–	–
6	1	25.80	33.33
7	1	19.35	25.00
8	1	32.25	41.67
9	–	–	–
Red loam soil			
4	–		
5	–		
6	–		
7	–		
8	2	83.84	
9	–		

TABLE 1.8: Level of Significance of the role of Forest with soil in Plateau Districts

Soil types	χ^2	Level of significance	Standard deviation
Laterites	11.48	0.01	404.17
Red and brown	9.29	0.001	316.46
Sandy loam	–	–	5.27
Red loam	–	–	–

The results show higher level of significance in laterite soil in comparison to (Table 1.8) Red and Brown soil with forest. However,

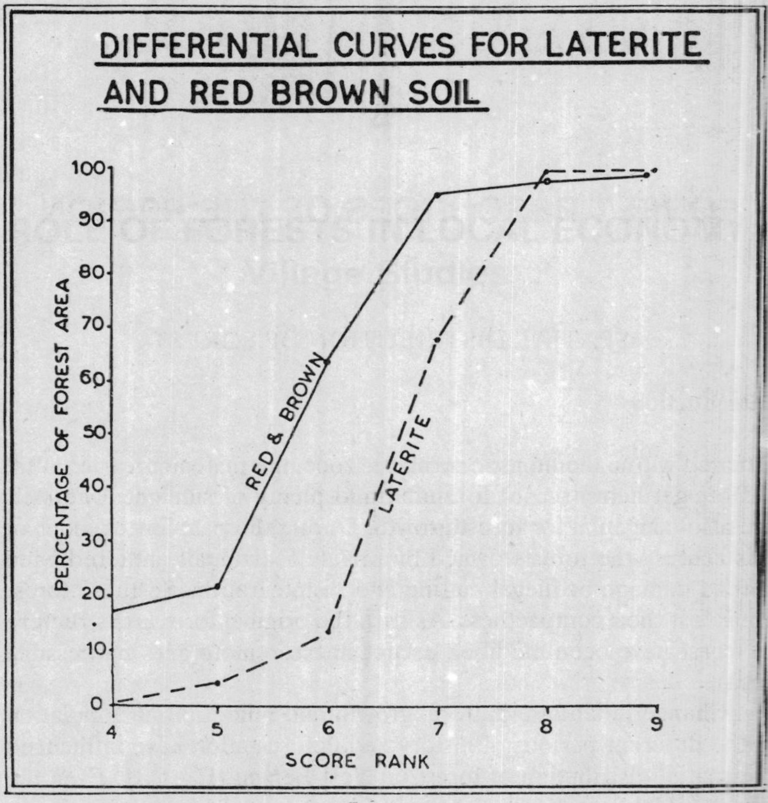

Fig. 1.8.

there is no significant relation of sandy loam or Red Loam with spatial distribution of forests in the Region.

Taking the cumulative frequency of the percentage of forest in laterite and Red and Brown Soil by the score values, a graph has been prepared (Fig 1.8). From the graph it has been found that about 97 per cent of the forest area occur in laterite soil, where the average annual rainfall and relief and lower than those in Red and Brown soil areas. Only three per cent of the forest area occur in Red and Brown Soil where both the rainfall and relief are low.

2

FOREST RESOURCES OF THE REGION

SPATIAL DISTRIBUTION OF FOREST

Introduction

Situated within the monsoon climatic zone, the plateau areas of West Bengal get heavy seasonal rainfall and plenty of sunlight. The soils are also congenial for forest growth. During the past few decades of this century the forests owned by private individuals, suffered wide spread damage by illegal cutting and disintegration. So these forest areas lost their compactness. As such the original form and structure of forest have been modified except in the remote and inaccessible areas.

Change in landuse pattern, growth and migration of population in the different periods of history and deforestation have influenced the spatial distribution of forest in West Bengal (Fig. 2.1). Even the forests which have managed to survive after prolonged exploitation, show signs of degeneration. Moreover, gradual progress of agriculture on the forest lands have enhanced the problem and change the residual vegetation to scrubs. Yet these forests provide the major-requirement of fuel, Fodder and constructional materials for thousands of people living in this areas.

Decadal Variations

During the earlier part of the 19th century, the forests taken together was known as Junglemahal. In 1833, this tract was divided between Birbhum, Midnapur, Bankura and Burdwan districts. In 1956 Purulia district was created after transfer of a part of Manbhum district of

Bihar to West Bengal (Table 2.1). As such the areas have also recorded changes.

TABLE 2.1 : Decadal Variations of Forest Area in the Plateau Districts of West Bengal, 1901-91

(Area in sq. km)

Year	Midnapur	Bankura	Purulia	Burdwan	Birbhum
1901	2590	1344	—	158	155
1911	2556	1344	—	158	155
1921	2911	1344	—	158	155
1931	2163	1344	—	158	129
1941	2028	1344	—	158	129
1951	2028	1401	—	158	137
1961	1684	1402	881	227	137
1971	1721	1395	876	277	157
1981	1708	1482	876	277	159
1991	1709	1482	876	277	159

Source : West Bengal forests, centenary commemoration volume. Government of West Bengal, 1964, West Bengal Forest Statistics, 1977 and West Bengal Forest Statistics, 1983.

TABLE 2.2 : Distribution Pattern of Forests in the Plateau Districts of West Bengal, 1991

(Area in sq. km)

Name of the districts	Geographical area	Forest area	% of forest to total area	% of forest to total forest area of the region
Midnapur	14,081	1,709	12.14	37.95
Bankura	6,882	1,482	21.53	32.91
Purulia	6,259	876	14.00	19.45
Burdwan	7,024	277	3.94	6.15
Birbhum	4,545	159	3.50	3.54
Total	41,587	4,503		100.00

Present Parttern of Distribution

The forests of the Plateau districts are characterized by the lack of compactness (Table 2.2). The upland ridges and hill slopes have thick forest cover while the low lying areas have patches of vegetation. In general, forests are found only sheltered patches in remote areas which are the remnant of the original forest cover of the Division (Fig. 2.1).

Bankura

In Bankura forests covers about one-fifth of the total area of the district. The forests form fairly compact zones in south and east. The important centres are Bishnupur, Ranibandh, Taldangra, Gangaalghati and Sonamukhi police station.

Purulia

The forests in Purulia are particularly concentrated in two centres : (i) the western part of Bagmundi police station near Ajodhya hills, and (ii) around Bundwan on the south-east. The forest covers 36 per cent of Bundwan police station.

Midnapur

Though in absolute terms the area under forests in Midnapur district is quite high, it covers only about 13 per cent of the total area due to massive size of the district. Thick patches of forests occur on the west and north, but the eastern side is almost devoid of it. The main concentrations are at Binpur, Jhargram, Nayagram, Salboni and Garbeta police station. The name Salboni is associated with sal trees, where forests cover about 38 per cent of the total area of this police station.

Burdwan

In Burdwan, forest area was formerly concentrated at Durgapur, Kaksa, Ausgram police station. At present the forest area has been considerably depleted with the installation of steel plant and other industry at Durgapur. As a result, only 4.4 per cent of the area of this district is under forests. Kaksa and Ausgram police station have considerable areas under forests.

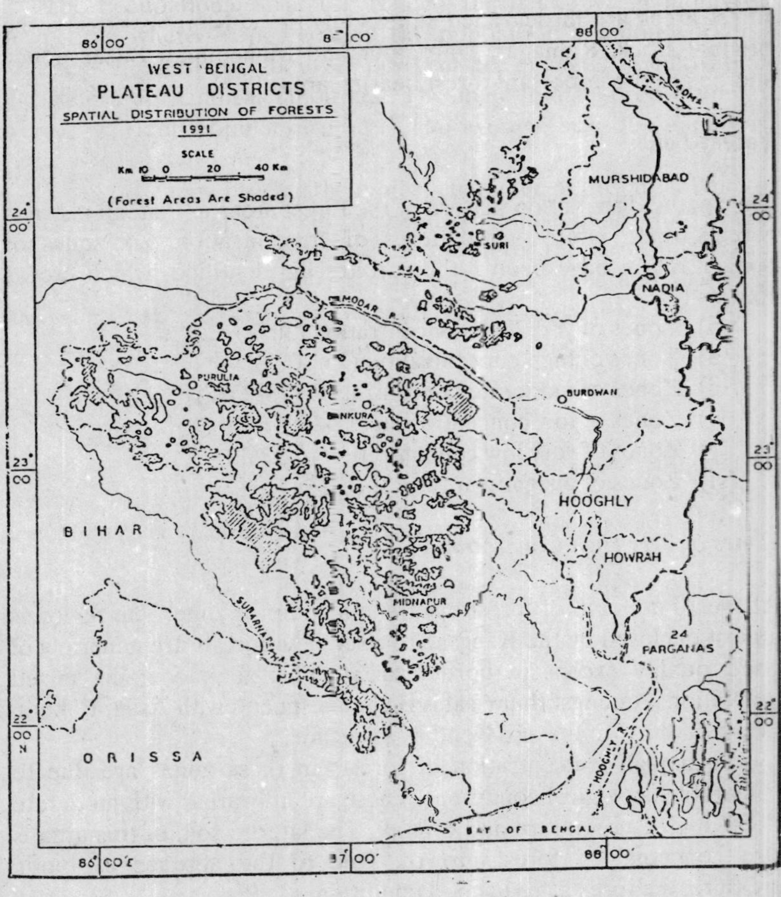

Fig. 2.1: Plateau Districts: Spatial Distribution of Forests

Birbhum

Birbhum district depicts a miserable picture. Most of the original forest areas are in degraded state. Only the core areas possess the original stock. Rajnagar, Dubrajpur and Mahammad Bazar police station are the important forest bearing area.

Forest Zones

The spatial distribution pattern of the forest areas in Plateau districts reveals some interesting patterns of concentration. Six zones of concentration have been rather arbitrarily identified, which are as follows :

(i) Zones of very high concentration (30%)
(ii) Zones of high concentration (25-30%)
(iii) Zones of moderate concentration (18-25%)
(iv) Zones of low concentration (12-18%)
(v) Zones of very low concentration (6-12%)
(vi) Zones of insignificant forest (6%)

Zones of Very High Concentration

Above 30 per cent of forest occurs in a compact zone. The region is almost enclosed by the Kangsabati river. The forests are generally of good quality except in border areas. This zone contains mixed vegetation. Amongst them, sal is the main species with Asan, Bahera, Dhaw, Sidha, Mohua and Kend as associates.

Such high concentration of forests in these zones are due to favourable ecological conditions. These are hilly areas with moderate slope, high temperature and rainfall. The laterite soil, of this area is also congenial to forest growth. Due to the absence of biotic pressure, the forests are in good condition.

Zones of High Concentration

This zone covers wide areas of Bankura and Midnapur districts where 24-30 per cent of the total area is under forest cover. Within this zone considerable forest area is man-made. Nayagram police station on the west of the Subarnarekha river also contains dense forest, which is virtually an extension of the forest areas of Orissa. In this zone, Sal is the main natural species. While eucalyptus and Akashmoni are the predominant planted species.

Zones of Moderate Concentration

This group claims 18-24 per cent of forest to total area. This is the region of miscellaneous species. Wood as well as timber production of this area continue to be very low.

Zones of Low Concentration

The concentration of forests is indeed low (12-18 per cent of the total area) at the edges of laterite tracts. In these areas, the predominant species is coppice sal of poor quality. This is mixed with kend, sidha etc.

Zones of Very Low Concentration

About two-thirds of Birbhum most of Hooghly, parts of Bankura, Midnapur and Burdwan districts belong to this group. In this group forests cover 6-12 per cent of the total land area.

Zones of Insignificant Forest

The eastern and south-eastern parts of Plateau districts have minimum forest area. However, there are some orchards and bamboo groves. The alluvial soils of the region hinder forest growth. As such the region has become excellent farm lands.

Types of Forests

As per the classification of Champion and Seth (1968), the forests of this region belong to Northern Tropical Dry and Moist Deciduous Forest Zones. The dry deciduous zone occurs in the Western part of the Plateau districts. The moist deciduous type, on the other hand, occurs at the eastern part. This zone again can be sub-divided into five types viz. (i) moist mixed deciduous, (ii) dry sal forest, (iii) moist low level peninsular sal, (iv) dry deciduous scrub, and (v) riparian vegetation.

Tropical Dry Deciduous Type

In Northern Tropical Dry Deciduous type, there is considerable intermixture of trees. A thin undergrowth of scrubs is found along with evergreen xerophytic species. An important feature of this type is the change in the appearance of vegetation with the change of seasons. In hot dry weather condition, the trees are entirely leafless and the soil becomes exposed, but in the monsoons with high rainfall the vegetation becomes luxuriant.

Within this group, *Dry Peninsular Sal* forest may be said to be a sub-type.The laterite tracts of plateau districts is covered with this type. This forest is usually found on shallow soils derived usually from crystalline and metamorphic rocks where the soil moisture is low. The main species are Sal, Bahera, Asan, Pial, Peasal, Simul, and so on.

TABLE 2.3 Forset Types by Classes (dia.)

(as percentage to total forest area)

Division	Below 10cm.	10-20cm.	20-30cm.	Above 30cm.	Mixed size
Sal forest area (in cm)					
Purulia	62.05	25.00	10.00	-	2.5
Bankura	62.50	23.75	12.50	1.25	-
Midnapur (E)	90.59	9.08	0.29	0.04	-
Midnapur (W)	83.45	16.18	0.31	0.06	-
Miscellaneous Forest area (in cm.)					
Purulia	31.60	34.20	5.30	2.60	26.30
Bankura	—	45.45	45.45	4.10	5.00
Midnapur(E)	NA	NA	NA	NA	NA
Midnapur(W)	NA	NA	NA	NA	NA
Plantation forest area (in cm.)					
Purulia	55.20	40.30	—	4.50	
Bankura	85.71	14.29	—	—	
Midnapur (E)	87.95	10.84	1.21	—	
Midnapur (W)	40,00	51.11	2.22	6.67	

NA = Not Available

Source : Forest Survey of India)

Tropical Moist Deciduous Type

This type of forest occurs in the eastern part of the districts. The forests are of irregular size consisting of tall and dwarf trees and are predominantly of deciduous type. The trees have wide girths. Some of them are of evergreen type. The floor is covered with undergrowth of shrubs. There are patches of bamboo groves as well. The orchards such as fig, mango etc. are common. The forest at this region is largely influenced by human activities. The sub-types of this variety are:

(i) *Moist Mixed Deciduous Forest:* This type of forest covers considerable areas mostly on the slopes of the main hills and accounts for nearly 40-45 per cent of the total hill area under this type. Floristically, this area is extremely complex and heterogeneous with poor percentage of valuable species.

(ii) *Dry Sal Forest:* The forest of this type occurs on the flat terrain with shallow soils. The area was noted for fairly dense sal forest in the past. But due to over exploitation, this type has now been replaced by xerophytic vegetation. At present, sal grows only on favourable sites where the soil character and moisture conditions are optimum for its growth. The quality of this species is poor. Natural regeneration of sal is sparse and its establishment is equally slow and uncertain.

(iii) *Moist Low-level Peninsular Sal*: Depending on the habitat factors, it is possible to distinguish three predominant facies within moist low level peninsular sal vegetation.

In the dry localities, *Sal Anogoissus* and *Sal Augeinia* communities are conspicuous. In moderately dry facies, *Sal Terminalia tomentosa* or locally *Sal Buchanania* communities are more frequent. In moist localities *Sal Terminalia tomentosa* communities are most frequent, giving place to *Sal Syzygium Rondia* communities in still moist localities near the streams and low lying areas. These communities occur in areas with deep well drained and fertile soil. Bahera, Hollandi, Mundi, Sidha are also fairly common associates.

These forests may, therefore, be described as climatic climax. In spite of a series of secondary succession which have been continuously taking place within the habitat, the sal forests of this type are more or less stable communities and are undistrupted by the climax vegetation.

(iv) *Dry Deciduous Scrub Forest*: In many areas, the natural vegetation has now been virtually reduced to scrubs under the impact of adverse biotic conditions. The present vegetation now consists of hardly species that can withstand hacking and grazing. The species generally found in such forests rarely attain the height of more than two metres and consist of shrubs such as, *kosee, jhatika, kendu, bheru* etc. The soil is shallow and compact with frequent limestone and laterite nodules. Thus the forests of these regions represent particularly the dry and floristically impoverished facies of the dry mixed deciduous forests where continuous biotic exploitation brought secondary succession and further degeneration of the plant community.

(v) *Riparian Forest:* This type occurs sporadically along the banks of perennial streams where the soil is sandy to sandy loam type and have abundant moisture. The principal species found are *arjun* and *hijal* (*Barringtonia accutangula)* and shrubs like Calotropis, Ipomeabiloba, Spinifax etc.

Composition of Forests

It has become quite apparent that long term impact of climate and soils have left marked influence on the evolution of the present day flora of this area. The forests of this area are primarily of recent type. Much of the earlier stands have already been exploited. The present day distribution of vegetation is thus merely the end product of the long evolutionary process of forest growth.

During the past, the forests of the area were subject to every conceivable form of mal-treatment such as reckless felling, working on too short a coppice rotation, lopping, fire and removal of stumps for charcoal production etc. Biotic activities in the shape of ground fire in summer, cattle grazing throughout the year and sweeping of forest floor for collection of leaves are common. Therefore, the forests have been reduced to low coppice type, and at many places it has become quite open. As such coppice forests cover almost the entire area with some exception in comparatively inaccessible areas.

The undergrowth consists of shrubs such as *Hollarhena antidysentrica, Gardenia qummifera, G. turgida, Croton oblongiforious, Flucourtia* spp. *Lantana* spp. *Zizyphus zuzuba, Combretum decandrus, Diospyrous melanoxylon, Acacia Pinata* etc.

Of the climbers *Butea superba, Dioscorea* spp., *Porana* spp., *Visits assamic, Melothria heterophylla, Spatholobus muenna, Abrus Precatorious* and *Milletia* are common.

The composition of forest occasionally depicts the climate of the area. For example herbaceous species indicates moist climate condition which are conducive to better natural regeneration of the species. On the other hand, occurrence of bel, kend and grasses is synonymous with definitely dry habitats where sal regeneration is scanty, slow and uncertain. The natural reproduction is virtually absent in localities where *Diospyrous peretrica* occurred to sizeable quantity. Some species are, however, associated with socio-cultural elements, such as mahua, *combrethus* shrubs, palash, asan etc. This is because the people can get revenue and food directly from them. They use mohus seeds for drink and mahua flower as food. Palash is used for lac cultivation. Asan and Arjun leaves are fed to the tassar silk worms.

According to composition, the forests can be grouped into three predominant types: (i) sal, (ii) miscellaneous, and (iii) plantation (Table 2.4). The forest areas of the plateau districts cover mainly Midnapur, Bankura and Purulia districts. The forests of Birbhum and Burdwan have less economic importance.

TABLE 2.4 Composition of Forest Types in Plateau Districts of West Bengal

(% to total forest area)

Types	Midnapur %	(Vol m^3/ha)	Bankura %	(Vol m^3/ha)	Purulia %	(Vol m^3/ha)
Sal	49.12	14.753	56.34	14.636	37.00	20.045
Miscell-aneous	35.80	–	23.94	22.025	36.00	25.280
Plantation	15.08	3.976	19.72	4.335	27.00	7.497
	100.00		100.00		100.00	

Source : Forest Survey of India, 1983

Sal : It may be of interest to note that sal forest covers the maximum space of the forest area in all districts. In Midnapur and Bankura, it covers half of the forest area. At places, this figure may even exceed 80 per cent to total forest area as in case of western Midnapur. In Purulia, sal forest covers less area because of its mixture with other species.

Though sal covers the highest area but the quality is not good everywhere. According to the quality it may be termed as nature sal,

Coppice sal and degraded sal forest.

Miscellaneous Forest : Although sal is the predominant species, miscellaneous species also occur in pure patches. Frequency of occurrence of miscellaneous species is more common in West Midnapur and Purulia districts. In West Midnapur Kusum are gregariously found. Bamboo is found all over the area. Among the miscellaneous species, asan covers the highest forest area with 51.061 stems/ha. Here the percentage of sal varies from 20 to 30 and the rest consists of the following species : Peassal, Dhaw, Mohua, Sidha etc.

Plantation Forest : Regarding the composition of plantation, it is found that Purulia possesses the highest percentage of plantation forest area (27%), being followed by Bankura (19.72%) and Midnapur (15.08%). Though plantation started in the middle of this century, the progress not satisfactory. Amongst the species planted, eucalyptus hybrid constitutes the principal species (81%). This species is worked under coppice system by clear felling having no shelter wood or mother tree. The density of tree reaches upto 830 stema/ha due to good locational factor. Acacia comes next. This species does not coppice on felling and consequently can be regenerated either by replanting or by natural regeneration in good areas. Small patches of plantation of sal, mahua, kend, chhatim etc. are found with specific objectives.

TABLE 2.5 : Statistical Data Base Midnapur District

Class	Area In Sq. Km	Percentage to the Geographical Area(%)
Dense Sal	438.154	3.11
Dense Mixed Forest	58.873	0.41
Open Sal	1713.956	12.17
Open Mixed Forest	203.285	1.44
Misc. Species	300.673	2.13
Grasslands	13.044	0.99
Degraded	468.854	3.33
Agriculture	7737.835	54.95
Wastelands	1300.66	9.23
Water and Build-up	1489.286	10.57
Cultural Features	64.269	0.38
Unclassified	292.154	2.07
Total	14081.00	

TABLE 2.6 Statistical Data Base Bankura Forest Division

Class	Area in sq. km	Percentage
Dense Sal	261.10	3.79
Dense Mixed Forest	109.46	1.60
Open Sal	916.35	13.31
Open Mixed Forest	380.27	5.52
Eucalyptus	29.60	0.43
Degraded	648.33	9.42
Wastelands	976.09	14.20
Agriculture	3014.48	43.80
Water	270.58	3.93
Sand	97.94	1.42
Unclassified	82.98	1.20
Cultural Features	95.21	1.38
Total	6882	

TABLE 2.7 Purulia Forest Division

Class	Area In sq. km	Percentage
Dense Sal	33.091	0.53
Dense Mixed Forest	136.294	2.18
Open Sal	318.025	5.08
Open Mixed Forest	739.512	11.81
Mining	3.818	0.06
Degraded	660.952	10.56
Wastelands	1469.131	23.47
Agriculture	2524.485	40.34
Water	157.551	2.52
Sand	27.983	0.45
Unclassified	94.618	1.51
Culture Features (Build up : 52.882, Rails : 9.828, Range Bdy; 11.384 and Roads : 19.349)	93.443	1.49
Total	6259	

Source : Deparment of Space, Regional Remote Sensing Service Centre, Kharagpore.

Conditions of the Forests

The laterrite section of West Bengal contains fairly large forest areas.

Agriculture was introduced here without any prior planning and control. As a result, the tract bears distinct evidence of moderate to extreme losses of soil moisture and complete disappearance of humus. The forest exists as enclaves or boundaries of agricultural field. Due to increasing demand in land for farming, forest lands had to be cleared. This in its trail brings other evils. Again due to such human action, the forests of yesterday have now been reduced to a degraded state. As a result, remnants of former luxuriant forests are only found today. In some places, one can see bare hills from which not only the vegetation has disappeared, but the soil has also been washed away. Lack of vegetation cover has exposed the soil to the direct action of the sun, wind, rain, thereby depleting its nutrients, so as to render it incapable of supporting even a field crop. Nature herself could make necessary recovery had these bare spaces been left to herself. But the action of man continues. Excessive grazing has rendered them totally unproductive. From the landsat imegeries, it has been found that the forest area decrease to a great amount which now covers only 10 per cent of the total area. The previous patches of forests are somewhere stand as remnants of the past or turn into a complete open land.

Height of the Trees

The height of the trees of the region records variations from hills to the plains. But if we consider the height of the trees individually, the height is more on the plains than hills because of the better soil texture and moisture holding capacity in the plains.

The top canopy of the forest in plateau section is 10-15m comprises of sal with mixture of miscellaneous species as Peasal, Bahera, Dhaw, Parashi etc.

The average height in the middle canopy is 8m and diameter varies from 5 to 20cm depending on the species. The density of vegetation varies with degree of human interference. The density in the top canopy is about 1,000 trees/ha and in the middle canopy about 2,000 trees/ha.

Forest Types according to their Nature

Depending on their nature, the forest areas of the region are classified into (i) coppice forest, (ii) jhanti forest, (iii) plantation forest, (iv) blank and scrub area.

Coppice Forest

Coppice forest covers bulk of the area of the forest, which at places covers up to 90 per cent to total forest area. As the main species is sal, it is called as 'sal coppice forest'. The quality of crop and ground vegetation indicate more strongly its similarity to dry sal type. The trees are dwarf with bushy growth. Sal is mainly coppiced because of its gregariousness and it ensures cent per cent replacement of older crop. Other species could hardly compete with sal in this respect.

Jhanti Forest

Near the towns or densely populated villages, fairly large blocks of xerophilous type of forests are found to have a dense shrubby growth with practically no wood. These forests are commonly known as 'Jhanti forest' or *Jhanti Jangal* in local terminology. These are the remnants of old sal coppice crops, Kurchi, Atari, Kul alongwith stunted Kend, bushy Parashi and scattered stools of sal.

Plantation Forest

In the degraded and blank areas forests have been created artificially by the forest division. The variation of species depend on the soil, climate and local need. Though plantation started during the middle of this century, plantation of quick growing species was started from 1962. The main species tried on a large scale are teak, acacia, cassia, eucalyptus, gammar, sal etc. Each of the species have auto-ecological character which are more or less known and are suitable for this region, though some of them are less effective.

Blank and Scrub Area

Small areas of scrub jungle of thorny species such as *Zizyphus flacourtia, Rhamnus gardenia* and as per are found in association with dry deciduous species. This is found on hard laterite, probably exposed, denuded hardened by human interference. This corresponds more or less with Champion's thorn forest of dry laterite zone. Significant areas of this forest type are the outcome of maltreatment of natural forests by man. In some ranges of Purulia and Bankura, about half of the areas belong to this type. About 44.66 per cent and 28.03 per cent of area of East Midnapupr and West Midnapur divisions respectively belong to open scrub forest category. Birbhum also have a higher percentage of area under this type. Thus with plantation, the forest area of there regions can be increased.

FOREST PRODUCTS

Forest Products as per their Characters

According to the characters, the forest products can be grouped into two categories viz., (i) major forest products, and (ii) minor forest products. The major forest product includes timber, pole fuelwood, wood pulp etc. The minor forest products includes sal seeds, Mohua seeds, Kendu leaf, lac, tassar and many other items.

The revenue from the forests is obtained from the extraction and yield value of these forest products. The forests of this region and however noted for the production of fuelwoods and minor forest products. These two items together account for nearly three-fourths of the total income of the forests of this region.

Systems of Sale

The disposal of forest products is done by auction, tender allotment and permit system. The timber and poles are generally sold through auctions. Pulpwood is supplied mainly on allotment basis to paper mills on receipt of royalty. The minor forest produce used to be sold through auction or permit system in past. Now LAMPS (Large sized multipurpose Society, affiliated to the Tribal Development Corporation) collect sal seeds, kendu leaves, sal leaves, mohus seeds on monopoly basis on payment of royalty.

Exploitation of Forest Products

Methods of Exploitation

Exploitation method is still primitive. This is mainly done by labourers of the purchaser. Among the species, sal trees are mainly felled, leaving a stump of not more than 15cm. on the ground. In case of Eucalyptus and Akashmoni 6 cm. of the stump should be on the ground. The picture of exploitation can be viewed from average outturn from Midnapur district. (Table 2.8).

TABLE 2.8 Average Outturn from Forest in Midnapur District

West Midnapur Division				
1.	High Volume area	– 1.3%	average outturn	30m^3/ha
2.	Medium Volume area	– 25%	average outturn	30m^3/ha
3.	Low Volume area	– 62%	average outturn	No yield
East Midnapur Division				
1.	High Volume area	– 10%	average outturn	52m^3/ha
2.	Medium Volume area	– 15%	average outturn	27m^3/ha
3.	Low Volume area	– 75%	average outturn	insignificant

Source : F.S.I., 1983

The low volume area covers about one-third of the forest areas, where the average outturn is insignificant. Sustained yield can therefore be expected only from medium to high volume area of the forests. The areas with poor density and damaged crops are not expected to provide any yield in the near future. Yet based on the aforesaid assumption, the average outturn of Midnapur district may be estimated as give in Table 2.9.

Considering Table 2.9 it can be said that though the overall picture of the forest products in the plateau district is discouraging, yet the area has promising prospects of regeneration if measures are taken for the protection. This is particularly so, because the production capacity of high volume and medium volume areas are quite satisfactory.

Major Products – Timber and Poles

Due to scarcity of large trees, production of timber is negligible. Most of them are below 10 cm in diameter. As such most of these products are treated as poles. The poles are mainly found from sal, peasal, mango, asan, neem, arjun, teak, siris, joral and mohua trees. Majority of the poles are sold in the range of 6-12 cm diameter at breast height (d.b.h.) in miscellaneous forests and upto 5 cm. d.b.h. in case of sal forests.

TABLE 2.9 : Production of High and Medium Volume Forest Area Midnapur District

(yield/ha)

		West Minapur	East Minapur
	High Volume Area		
1.	Area of Sal Stratum	70,667.80 ha	7,5531.00 ha
2.	Extent of high volume area 13 per cent for West Mid. 10 per cent for East Midnapur.	9,187. 00 ha	7,553.00 ha
3.	Extent of workable area (95% of high vol. area 5% of the area is kept as reserved for silvicultural and technical reason)	8,728.00 ha	7,175.00 ha
4.	Rotation	12 years	10 years
5.	Annual cut area	72 7ha	717 ha
6.	Vol/ha.	30 m^2/ha	52 m^3/ha
7.	Annual yield	21810	37284 m^3
Medium Volume Area			
1	Medium Vol. area 25 per cent for West Mid. and 15 per cent for East Mid.	17667	11330
2	Extent of Workable Area (95%)	16784	10763
3	Rotation	12 years	10 years
4	Vol/ha	30 m^3	27 m^3
5	Annual Yield	4197 m^3	29052 m^3

Source : F.S.I., 1983.

Poles of other species such as dhaw, pial, kend etc. have also good market value. Some common timber which are found at large girth rate are simul (60 cm), sirish, haldu etc.

The low production of timber and pole may be ascribed to the following factors :

(i) It has been found that approximately after 15-16 years, the sal species usually get infected with fungus, mainly on their heart wood. So before being attacked by the fungus, they have to be cut. Thus the optimum girth needed for timber is rarely attained; and

(ii) Unauthorised felling is carried all over the area where forests are not allowed to grow to its maximum volume.

Fire Wood

Fire wood is considered to be the most important resource and highest revenue earner of the forest of this region (Fig. 2.2). The trees below 5 cm. diameter is used as fire-wood.

The sale of fire wood is done through auctions. The State Forest Department disposes off the coppice-coupe lots through auctions. Timbers, poles and fire woods are all included in that coupe. It has been estimated that about 70 per cent of the total stems have 5-10 cm girth rate. This naturally results in high production of fire wood. The outturn of fire wood is about 54.9 m^3/ha, though it varies from division to division and year to year (Table 2.10 and 2.11).

TABLE 2.10 : Division-wise Outturn of Forest Products (Forest Directorate and Forest Corporation)

(in cubic metre)

Division	1985-86		1986-87		1987-88		1988-89	
	Timber & Poles	Fire-wood	Timber & Poles	Fire-wood	Timber & Poles	Fire-wood	Timber & Poles	Fire-wood
Bankura(N)	10987	25918	354	3679	344	1394	8625	291
Bankura(S)	1410	5379	2196	409	27	163	4919	1307
East Midnapur	7734	3748	2672	11498	2411	20004	8844	2626
West Midnapur	1004	8834	649	5852	1191	10120	1301	5015
Burdwan	177	4658	986	2207	581	1379	632	4431
Purulia	694	502	422	647 *	—	—	—	—
Birbhum	667	4420	728	3215	60	991	109	556

*No clear felling

Source : Forest Directorate, Government of West Bengal.

Pulp Wood

The production of pulpwood was started with the initiation of afforestation programme in 1947. But the real production was initiated during 1962 with the plantation of eucalyptus hybrid. Eucalyptus and *Accacia auriculiformis* are mainly used for pulpwood

TABLE 2.11 : Year-wise Revenue from Major Heads

(*In lakhs*)

Divisions	Timber	Fire-wood	Minor forest Produce	Others	Total
1	2	3	4	5	6
1985-86					
Bankura(N)	7.08	76.80	1.00	4.24	89.12
Bankura(S)	1.74	15.15	0.27	4.52	21.68
E. Midnapur	9.44	98.74	2.53	6.10	116.81
W. Midnapur	0.41	42.04	0.66	4.93	38.04
Burdwan	2.15	1.43	0.36	0.86	4.80
Purulia	3.25	0.40	0.40	1.96	6.01
Birbhum	0.73	6.73	0.07	2.32	9.85
1986-87					
Bankura(N)	8.11	10.97	1.18	3.58	23.84
Bankura(S)	11.63	1.75	0.47	2.01	15.86
East Midnapur	2.24	36.32	7.97	4.62	51.15
West Midnapur	0.73	21.06	2.97	3.40	28.16
Burdwan	1.10	0.18	0.27	1.64	5.89
Purulia	3.67	0.31	0.27	1.64	5.89
Birbhum	1.45	3.83.	0.14	2.26	7.68
1987-88					
Bankura(N)	0.35	11.88	1.07	1.72	15.02
Bankura(S)	6.39	0.39	0.74	1.94	9.46
East Midnapur	7.81	7.44	11.05	0.18	26.48
West Midnapur	0.35	8.13	3.41	3.10	14.99
Burdwan	3.80	1.30	0.48	0.76	6.34
Purulia	3.09	0.07	0.57	1.45	5.18
Birbhum	1.09	4.18	0.25	2.25	7.77
1988-89					
Bankura(N)	0.02	5.62	0.75	1.31	7.70
Bankura(S)	9.92	0.29	0.76	2.16	13.13
East Midnapur	4.27	1.82	4.57	3.58	14.24
West Midnapur	1.19	4.86	2.51	3.97	12.53
Burdwan	3.08	1.54	0.43	4.66	9.71
Purulia	1.79	0.05	0.44	1.08	3.36
Birbhum	2.15	1.38	0.07	1.73	5.33

Divisions	Timber	Fire-wood	Minor forest Produce	Others	Total
1989-90					
Bankura(North)	0.22	12.51	1.10	2.05	15.88
Bankura(South)	5.10	0.15	0.99	2.67	8.91
East Midnapur	57.49	13.99	9.37	5.25	86.10
West Midnapur	8.35	17.51	2.40	10.28	38.54
Burdwan	9.47	4.79	0.40	3.81	18.47
Purulia	2.08	0.05	0.42	1.45	4.00
Birbhum	2.03	0.06	0.10	3.79	6.52

Source : Forest Directorate, Government of West Bengal.

production. The consumption of pulp wood newsprint, paper and paper boards are steadily increasing since independence following the growth of literacy in the country. So there is a corresponding increase in the demand of pulp woods in the market. Eucalyptus and akashmoni presently cover about 62.7 per cent and 26.7 per cent of the total area in this region respectively. The figure varies depending on their suitability.

With this high percentage of plantation, this region can reasonably produce sufficient amount of pulp wood. But from Table 2.12 it has been found that only negligible percentage of the area worked out and average yield is very low.

TABLE 2.12 : Area worked out by the paper mills in the Plateau districts (1982-83)

District	Area worked out(ha)	% of area worked out	Outturn (tonne)	Average (tonne/ha)
Midnapur	1,304	3.09	26,960	20.67
Bankura	1,289	3.34	12,410	9.63
Purulia	2,181	4.93	19,252	8.83
Birbhum	324	6.22	324	1.00
Burdwan	112	1.23	1,524	13.61

Source : Same as Previous Table.

The highest yield is recorded from Midnapur. The area worked out by the paper mills and the out trun of pulp wood are very low

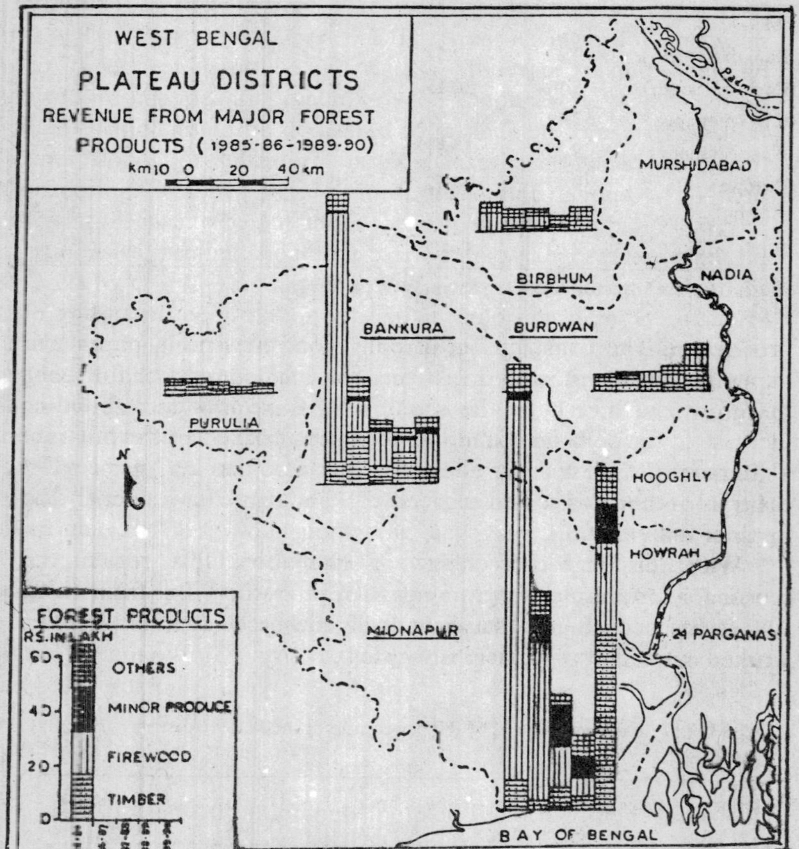

Fig. 2.1: Plateau District : Revenue from Major Forest Products (1985-86 – 1989-90)

because of the illegal poaching of trees at relatively young age. As such, majority of the crops fall below 10 cm diameter which hinder pulpwood industry.

The average volume of the species in plantation area is about 7m³/ha density is 167 stems/ha. But is some places of Bankura district, it has been found that the tree density reaches up to 680 stems/ha which indicates that with proper care the area will be able to produce more pulp wood.

Minor Forest Products

Continuous biotic interference and absence of production leads to the forests of the plateau section to a degraded state. Despite such poor state of affairs, these forests hold good promise in the yields of minor forest products. From the economic point of view, the value of minor forests products is not very high, but in upgrading the economic status of the poor villagers, these products play significant roles. About 10 to 15 per cent of the villagers of the region fully depend on the minor forest products for their sustenance. The minor forest products of this region are lac, tassar, chasewnut, sal seeds, mohua seeds, kend leaves, sal leaves, mohua flowers, fibres and flosses, grasses, fruits, barks, gums and resins, drugs and insecticides etc.

Though this area possesses substantial amount of minor products, the extent of their utilization remains poor. This may be due to the following causes :

(i) These minor products are considered less important from the view point of forest resource and national economy;
(ii) Little work is carried out to generate information for the better management of minor forest products; and
(iii) Lack of marketing facility and dominancy of middle man, act as deterrent factors for their efficient use.

Besides the economic considerations, other issues are the welfare of the tribals and their employment generation capacity. A large segment of people living in forest areas, are directly, and indirectly associated with the collection of these products. Thus if arrangements are made for their proper development, it will not only generate more employment with its long chain of production, but can also increase the earning capacity of the tribals as in case of lac, tassar, sal seed oil, where the length of chain can be extended to wider spheres of economy.

Management of Minor Forest Products

The task of management of minor forest products has been left unattended. Only *adhoc* and short-term arrangement have been made in the past. The task of their efficient utilization and marketing have now been entrusted to a state sponsored Large-sized Multipurpose Society (LAMP). The LAMP is collecting these products on payment of royalty to the government from 1977-78 at a fixed rate. The cashew cultivation has also been entrusted to the LAMP at the initial stage. But government has now assumed control over it. In many States, trading of these minor products have been nationalised as in case of kend leaves in Madhya Pradesh. But no such programmes has been taken up here. Such difference in organisational arrangements therefore create problems.

The method of collection and preparation of the finished products are often crude. In this process the quality of these products often deteriorate. The products are then sold to large dealers. The products have to pass through several hands before they are eventually brought to the ultimate consumers or exporters.

Owing to the scattered occurrence of the products, not only the cost of collection becomes high but the cost of transport is also increased. Again, there is great discrepancy between potential collection and actual collection of sal-seed, neem-seed and kend leaves.

The discrepancy may be for various causes. The physical factors which hinder production are :

(i) scattered growth of trees in this area;
(ii) wide variation in yield from year to year. Generally there is one good year for the production of sal-seeds in every three years; and
(iii) the perishable goods are very difficult to preserve.

The non-physical factors which hinder production are :

(i) The collection of these minor products are done through

LAMPS. So the efficiency of collections depends upon the organisational capabilities of each individual LAMPS. In some place, lack of activities or non-existence of such unit may result in failure of their collection;

(ii) Lack of proper marketing facilities; and

(iii) Absence of storage facilities.

The present method of collection and marketing are inadequate and any improvement in this direction would certainly lead to a considerable increase in the quantity collected and consequent benefit to local economy. In some cases, lack of sufficient information in the distribution, availability, method of collection, best season of collection and grading etc. have stood in their way of exploitation, as in case of lac, tassar, cashew etc. Though they are important revenue earners, but due to the lack of proper management and government patronage revenue earnings from them continue to remain low.

Oil Seeds

Increasing demand of vegetable oils in the country, has created demand on the minor oil seeds of tree origin. Though the oil obtained from the seeds of sal, mohus, kusum, karanj are not preferred like the common oils, but these have non-edible uses in the manufacturing of soaps and lubricants for leather industry and in medicine which will help to relieve pressure on edible oils.

Sal Seeds

Out of all the oil seeds, the seeds of sal come in the forefront due to its gregarious occurrence all over the area. The cotyledons contain 72 per cent of the weight of a seed and yield 19-20 per cent of fatty oil known as sal butter. The seeds are husked and boiled and the oil is skimmed off. The oil soon solidifies to a greenish white and whitish colour butter. It is used for cooking and lighting and may be mixed with ghee. It can also be used in soap making, leather polish and as a Cocoa butter in the manufacture of chocolate as coating.

For its versatile use, the demand of sal seeds is increasing day by day. But the production of sal seeds largely depends on age and density of the crop and periodicity of seeding. Generally good seed year occurs once in three years.

There was no commercial demand of sal seeds before 1973. The production also varies sharply from one place to another. The production of oil in western plateau is indeed low - 52 kg per ha, as against 152 kg in North Bengal.

The revenue (Rs. 1.50 per kg) earning is not up to expectation till now. But with proper care, more revenue can be earned which will also be beneficial for the local people.

Mahua Seeds (Modhuca latifolia)

The mahua are large deciduous trees found all over India. The seeds of mahua yield 20-30 per cent oil which is used in the manufacture of soaps, vanaspati, confectionary, candle etc. The oil extracted from the kernel is largely used as a substitute of cocoa butter. It has also some medicinal value.

Mahua trees starts flowering at the age of 10 years after it attain 20 cm in diameter and becomes mature for flowering and fruiting at 15 years. The tree has productive life span of 85 years. Mahua flowers are dried and consumed by the tribals as staple food during the lean period. It is an important fodder. Mahua flowers are used to make country liquor which is widely used by the tribals. The collection and marketing of Mahua seeds and flowers are done by the tribals from March to June.

It has been found that the production of seed is approximately 20 kg/tree. The occurrence of mahua is about 6.6 per cent in Bankura, 0.24 per cent in Midnapur East and 2.67 per cent in West Midnapur.

No detailed statistical study has so far been undertaken in this stage to estimate the potential production of mahua seeds. However, considering the density of one mahua tree per 10 hectares, the potential production of mahua seeds may be computed as 901 tonnes. This has been calculated on the basis of the mahua bearing area which is approximately 450,600 ha and seed production is 20 kgs. per tree. But the actual collection of mahua seeds is shown in Table 2.13.

TABLE 2.13 : Actual Collection of Mahua Seeds in West Bengal Plateau

Year	Production (Tonnes)
1977	40.79
1978	52.9
1979	43.84
1980	26.97
1981	27.0

Source : Directorate of Forests, West Bengal, 1983.

Thus there is considerable scope for maximisation of the yield of mahua seeds, as the difference between actual and potential collection is very high.

Neem (Azadirachta indicia)

Neem can be considered an indigenous tree associate of the forest areas having sufficient production. The oil is used for medicinal purpose and for soap making. About 200 kg of neem oil are needed to manufacture one tonne of soap. Due of absence of any government initiative, production of neem oil is poor. Collections during the past few years are shown in Table 2.14

TABLE 2.14 : Collection of Neem Seeds

Year	Production (in kg)
1977	105
1978	120
1979	145
1980	110
1981	123

Source : Forest Directorate, West Bengal 1984.

Palas (Butea prondosa)

Palas seed produces oil for soap industry. According to a recent survey conducted by the forest department of Bihar, the average yield of palas seeds is 0.79 kg per tree. In the absence of any such survey in West Bengal, the potential production of palas seeds is difficult to ascertain. In Purulia district, there are about 103,513 trees in the forest which can produce 82 tonnes of seeds. Thus from the forest areas of Bankura, Midnapur and Birbhum, more palas seeds can be collected. This will help to extract more oil beneficial for allied industries.

Kusum (Schleichera trijuga)

The oil extracted from kusum seeds is used for cooking as well as for lighting. It is also used in soap making and as hair oil and lubricant. It yields 25-27 per cent of oil. The fruits are consumed by the villagers during hot dry season. Now due to illegal cutting, the number of trees is gradually dwindling. But with protection and care, it can be increased and may be used as substitutes of valuable products. In Purulia district, 5,057 quintals of seeds and 59 quintals of fruits are harvested in a year. Moreover, palash and kusum plants act as host trees for lac. So along with the lac production, the villagers can also extract oil.

Karanj (Pongamia pinnata)

The karanj seeds produce 18 to 22 per cent of red brown oil which is used in lighting, in pharmaceutical and paints industry. It is also used as a medicine for skin disease.

Kendu Leaves (Diospyrus melanoxylon)

The kendu or biri leaves are in demand owing to their extensive use in the manufacturing of biri (indigenous cigarette). The collection is made from new plants because the new leaves are only used in biri. One female worker can collect 5-6 kg of leaves/day. The cost of collection of kendu leaves is also high due to scattered location of the plants. Now LAMPS have been entrusted with the right to handle kendu leaves for the State.

In West Bengal, kendu leaves are only found from the plateau districts. Among them, Bankura and Midnapur districts are the lending producers. Chirugara at Binpur P.S. Midnapur is noted for its cultivation (Table 2.15).

Compared to other States, the production of kendu leave in West Bengal is very poor though it has now assumed a great economic importance for its demand in the ever increasing biri industry. In Madhya Pradesh, Orissa and Andhra Pradesh, kendu leave have great commercial value and fetch substantial amount of revenue.

TABLE 2.15 : Collection of Sal Seeds and Kendu Leaves, 1988-89

Districts	Sal Seeds in M.T.	Kendu leave in Fali bags*
Bankura	427.21	54,771
Midnapur	530.67	25,880
Purulia	37.35	20,401
Burdwan	–	1,133
Birbhum	47.55	1,689
Total	1,042.78	1,03,874

*1 Fali bag = 20 kg
Source : Forest Directorate, West Bengal.

Future Planning

The following suggestions may be adopted for the improvement of kendu plantation :

(i) Systematic research on various aspects of the cultivation of kendu plant should be carried out with a view to increase the yield of leaves per unit area;
(ii) Collection of leaves in proper time should be done. There is a need for proper storage of leaves, so as to protect them against damp and insect attack. It is also necessary to improve the marketing system;
(iii) If the trade of kendu leaves is organised properly, it will improve considerably the economic status of the village people specially the tribals;
(iv) As a part of long-term programme, the kendu species should be planted in a compact block so that they could easily collect.

Sal leaf

Collection of sal leaves for making indigenous plates is another household industry in which women-folk are engaged. A large number of sal leaf plates are made at village level and sold in the markets. These plates are usually marketed in Calcutta and other towns. But collection of sal leaf is low paying occupation. The price for one bundle of sal leaf plate is Rs. 22 consisting of 1000 plates of 30 cm diameter. Price of 600 leaf plates of smaller size cost Rs. 3.50 only.

The cost of collection varies from time to time. It has been estimated that at Chondrakona Road in Midnapur the collectors usually get on an average Rs. 20 lakhs per annum. It is an important marketing centre and have links with other districts. Collection of sal leaf and making plates from them provide good employment to the rural people, both as full time or part time basis.

Lac

Lac in its refined form is popularly known as shellac. It owes its origin from the secretion of lac insect. It is therefore not a vegetable resin. Apart from the resin, a crimson dye is also derived from the lac insect. A lac host yield about 2 to 5 kg lac/tree.

Uses of Lac

Lac is one of most versatile natural resinous materials possessing a unique combination of various properties. Lac finds a wide variety of application in plastics, electrical adhesive, leather and wood finishing and in other industries. Before the invention of plastics, lac was used in gramophone. Surface coating industry is the largest consumer. Lac has long been in use both for decorative and insulative varnishes and lacquers of various kinds. It is used as a coat for metalware to prevent tarnishing and for finishing various other products, such as playing cards, oilcloth and linoleum. It is the principal ingredient of sealing wax. It is also found in glazing paper, nail polishes, dental plates, bangles, jewellery fittings, grinding wheels, pyrotechnic and confectionary.

The average stick lac contains about 68 per cent resin, 10 per cent dye and 6 per cent of waxy substance.

In West Bengal, lac is cultured in Purulia and Bankura district. The lac processing factories are located at Jhalda and Balarampur of Purulia. Superior quality lac is exported through Calcutta. More than 45 factories are operating which employ approximately 1,600 workers per day. About nine per cent of lac production in India comes from Purulia.

Tassar

Tassar is considered to be the most costly minor forest product used for silk production. The two important species Arjun and Asan are major host plants which are grown as the associates of the major plants in the plateau areas of Purulia, Bankura and Birbhum. The tassar culture is going on here for centuries. But now for steady demand in home market and abroad there is ample scope to segment its production. As such it is necessary to augment tassar production. In can be only possible through plantation of more Asan and Arjun trees and to feet their leaves to tassar-cocoons. Such an experiment has already been initiated at Chandrapura village in Birbhum district. The experiment has revealed the following results :

The cost incurred for the tassar cultivation for the two crops in one ha. of land : Rs. 1,400

Return from the Tassar cocoons : Rs. 2,000
Net Profit : Rs. 600

So the profit from the cultivation of tassar host plants and cocoons are Rs. 600. Along with this, for the process of rearing of Cocoons, 60 mandays are generated at the rate of Rs. 8.58 per mandays. For the reeling and weaving 60 days generate 120 mandays.

The cost and benefit of silf production form it :

(i)	Total cost incurred for the production of cocoons in 1 ha. land	Rs. 1,400.00
(ii)	Cost for weaving and reeling	Rs. 460.00
		Rs. 1,860.00
(iii)	Expected return from silk	Rs. 2, 350.00
(iv)	Net Profit	490.00

Source : K.C. Chokravorty, D.F.O. Birbhum, 1980

Though the production and profit is not high but with more attention 'and improvement in weaving and reeling it is possible to fetch more profit. The process can possibly be introduced all over the plateau area and will certainly help a substantial percentage of the tribal population of the area.

Cashew-nut (Anacardium occidentale)

Cashew nuts are probably the costliest of the numerous minor products obtained from the forests of this region. Because of its good taste, the nuts have great demand in market. Not only the cashew kernel but juice are also produced from Cashew apple and oil is extracted from its shell. Cashew-shell oil is used as raw material in lumination, brake linings, electrical installation, surface coatings, paintings etc., contains about 20 per cent of oil. Cashew cultivation is carried on in small scale in Midnapur and Bankura districts but not in systematic manner (Table 2.16).

TABLE 2.16 : Production of Cashew

(*in tonnes*)

Districts	1978-79	1979-80	1980-81	1981-82
Midnapur	80	199	126	–
Bankura	–	2.5	2.6	1.45

Source : Forest Survey of India, 1983.

Though the production is low, cost is very high due to high demand in the world market. The average market is Rs. 200 per kg in West Bengal today.

In West Bengal, Midnapur is the leading producer of cashew nuts. No large scale factory has yet been installed for processing the nuts, as well as for the recovery of its by products. Now under the social forestry programme, cashew is being cultivated as mixed crop.

As laterite morrum soil and the sandy loams of the region are congenial for cashew plantation, so it can be cultivated all over the plateau section. At the end of the Fourth Five Year Plan, the area under cashew cultivation was 2,308 ha. This has been increased to 3,200 ha during the Fifth Five Year Plan. It is highly profitable product.

If the high yielding variety cashew seeds are used in plantation, it will yield at least 20 kg of kernal per tree i.e., 20,000 kg. per 1000 plants. The total cost of plantation in the four years has been estimated at Rs. 1,600. The quantity of the yield will be atleast 1/4 of the total i.e., 5,000 kg. The value will be 5,000 kg. × Rs. 60 (@ Rs. 60 per kg.) Rs. 3,00,000 at the current price for each 1,000 plants. This clearly indicates that the programme is highly profitable. (Mukerjee, 1982)

But the production continues to be low due to various problems such as :

(i) Reliance on traditional varieties and poor productivity of these seeds;
(ii) Poor financial position of the growers;
(iii) Lack of proper technical knowledge; and
(iv) Uneconomic units of individual plantation.

The following measures may be adopted to alleviate the above problems:

(i) Strengthening the know-how;
(ii) Introduction of high yielding varieties of plants from Arabari forest department, Midnapur or from Andhra Pradesh;
(iii) Provisions of necessary finance to the growers at least during the initial phase of plantation;
(iv) Cultivation on economic holdings; and
(v) Cultivation of cashew as an inter-crop with Eucalyptus.

All these measures can be adopted through realistic planning staffs of the State Forest and Agricultural Department, involvement of political parties and traders are necessary. The crop has bright prospect in view of its geo-economic adaptability to this region.

Bamboo

With diverse and manifold use, bamboo can be cultivated in the degraded lands and on the adjoining home steads. One variety which is already experimented at Arabari, proves that one bamboo can produce 20/30 stalks of bamboo (as many nodes as the bamboo have) so it can be cultivated in small holdings. Previously farmers have to buy rhyzome which costs Rs. 10 per nursery plant. But due to scarcity of rhyzomes, they have to face difficulties. Now under new method, one bamboo which costs Rs. 20-25 could produce 20-30 bamboo and help to earn Rs. 400 to 750 at the rate of Rs. 20 to 25 for each stalk. For the increasing demand in paper industry bamboo have got an assured market.

Casuarina

Amongst the 20 species of casuarina, only 3 or 4 varieties are tried in India. It is a quick maturing tree, preferring loose fine sandy soil of the sea cosast, in hot climate, with moderate to heavy rainfall. It has been found that from this species chemical, semi-chemical and chemi-mechanical pulp can be prepared with satisfactory yield (Guha, *et al*). The strength of the pulp production is the same as that of *Eucalyptus grandis*. This is used in the production of wrapping papers, corrugating medium and core of duplex board.

As a proposal it can be said that the coastal area of Midnapur should be covered with casuarina plantation. The cost benefit analysis depicts a bright picture. The cost of formation of casuarina plantation has been estimated as Rs. 250 per acre in the first year, Rs. 100 in second year and Rs. 10 in each of the third and fourth year.

The yield of casuarina from 10 years old plantation, is 63.72m^3 fire wood and 275 number of poles from one acre of land. This fetches a revenue of Rs. 1,800 per acre from the 10th year onwards.* Thus it is possible to produce a substantial quantity of raw material for paper and pulp industry. It not only supplies the raw material needs of paper mills but helps in stabilizing the coastal sands.

Babui grass (Eulaliosis bimata)

The making of rope from Babui grass is another traditional household industry in this plateau. It employees women folk and children. The high lands are used for the production of Babui grass. The grasses are planted at an interval of 0.7 metre. About 4-5 bunches produce 1 kg of Babui rope. It is a supplementary occupation. Cultivation of Babui grass is prevalent in all districts but it is cultivated on substantial areas in Midnapur district.

Gums and Resins

Gums are collected from various trees such as :

* Surveyed in 1980.

Sherculia urens
Cochlospermum gossypium
Anogeissus latifolia
Bosswellia serrata
Lannea grandia
Buchanania latifolia
Terminalia belerica

The gums are in the manufacture of food articles, pharmaceuticals, cosmetics, polishes, adhesives, paints, varnishes, textiles, paper, explosives etc.

3

ROLE OF FORESTS IN REGIONAL ECONOMY : Regional Survey

OVERALL ECONOMIC BASE

Economically, the plateau districts have been demarcated as backward areas because of its insignificant industrial development and poor productivity of farm lands. The economy of the region is agro-based but the agricultural production is very low. The small nutrient content in the soil does not encourage agricultural development. The climatic, economic and social factors are not congenial for agricultural growth. The low productivity is also due to fragmentation of land, thereby rendering the land-holding very small and uneconomic. About 85 per cent of the total number of holdings are less than 0.5 ha in size (Table 3.1).

TABLE 3.1 : Pattern of landholdings in West Bengal

Size Class (ha)	Number ('000)	% of holdings in size class
Less than 0.5	15,558	85.41
0.5-1.0	971	5.33
1.0-2.0	942	5.17
2.0-3.0	412	2.26
3.0-4.0	145	0.80
4.0-5.0	98	0.54
5.0-10.0	87	0.48
10.0-20.0	3	0.01
20.0-50.0	0.1	—
50.0 and above	0.3	—

Source: Documents of the World Bank Report, Government of West Bengal, 1981.

The pressure of population on small holdings are particularly acute because of high density of population, with a corresponding dearth of good agricultural land and poor growth of industrial activities.

Midnapur, Bankura and Purulia districts are the important forest producing regions where more than 90 per cent of the total population are rural. Bulk of this rural population are engaged in primary economic activities, so the dependency rate is very high (Table 3.2). The percentage of non-working population vary from 60 to 70.

TABLE 3.2 : Rural and Urban Composition of Population, 1981

District	Area in Sq. km	Percentage of rural pop.	Percentage of urban pop.	Percentage of total pop. of West Bengal	Density of pop. per sq. km
Midnapur	14,081	91.50	8.50	12.35	479
Bankura	6,882	92.37	7.63	4.35	345
Purulia	6,259	91.00	9.00	3.40	296
Burdwan	7,024	70.61	29.39	8.86	688
Birbhum	4,545	91.72	8.28	3.84	461

Source : Final Population paper 1 of 1982 series, 23, West Bengal, Census of India, 1981.

The backwardness of this region is further enhanced by the high percentage of scheduled caste and scheduled tribes in the total population (Table 3.3).

TABLE 3.3 : Percentage of Scheduled Caste and Scheduled Tribe to the total Population, 1981

District	Total population	Total scheduled caste population	% age of total S.C. popu. of the district	Total S.T. population	% age of total S.T. population of the District
Midnapur	6,742,796	984,731	14.60	538,877	7.99
Bankura	2,374,815	686,300	28.90	250,590	10.55
Purulia	1,853,801	348,217	18.78	348,375	18.79
Burdwan	4,835,388	1,213,435	25.09	278,191	5.75
Birbhum	2,095,829	624,521	29.80	145,133	6.92

Source : Forest Statistics 1983.

The scheduled caste and scheduled tribe population rely to a considerable extent on forests to sustain themselves. So forests play a significant role in moulding their economy.

Livelihood Pattern

The overall economic structure of the people of the region is very poor. About one-fourth of the total population is classified as workers. Most of them are agricultural labourers and cultivators. The employment opportunities are not uniform throughout the year. Very little area is cultivated twice in a year. About 30 per cent of the population are living below poverty line.

TABLE 3.4 : Classification of Working Population by Broad Categories, 1981

Districts	Percentage of worker	Percentage of cultivators	Percentage of agri. lab	Percentage of household industry	Percentage of other workers	Percentage of marginal workers	Percentage of non-workers
Midnapur	33	15	8	1	6	3	67
Bankura	33	12	10	1	6	4	67
Purulia	39	16	8	1	7	7	61
Burdwan	33	6	8	1	13	5	67
Birbhum	31	11	11	1	5	3	69

Source : *Economic Review*, 1982-83, Government of West Bengal.

Though about 70 per cent of the population are directly linked up with agriculture, but in the plateau districts due to its low productivity, the crop can provide food to the inhabitants for only 3-4 months. The household industry supports only one per cent of the employed people. People engaged in other works claim 5 to 13 per cent of the total population.

Industrial Occupation

The number of registered small scale and cottage industries in this region present a poor picture. Actual employment or employment potentiality in this section is indeed limited.

TABLE 3.5 : Number of Small Scale Industrial Units and their Corresponding Employment in Plateau districts (1982)

Districts	No. of Units	No. of people employed	% age of total employed people
Midnapur	937	6915	0.10
Bankura	343	1920	0.80
Purulia	436	3158	0.17
Burdwan	1,297	9689	0.20
Birbhum	880	6662	0.32

Source : *Economic Review* 1982-83, Government of West Bengal.

Role of Forests on Economy

The forest areas of the plateau districts are mainly concentrated in Midnapur, Bankura and Purulia districts. Patches of forest also occur in Burdwan and Birbhum. Here in areas of substantially which forest cover, people are fully or partially dependent on it. They collect wood for domestic use as well as for sale to earn their livelihood. The rural people rely heavily on forests for meeting their day-to-day needs of fuelwood, fodder, agricultural implements, constructional materials etc. The life as well as the lore of the tribal people revolve greatly round the forests. Some of them fully depend on forests.

TABLE 3.6 : Percentage of Population Depending on Forests, in some Villages of Purulia District, 1983

Village	Wood cutting and other occupation	Wood cutting
Dandudih	40.01	2.86
Ghatber	64.14	14.49
Gittinglahar	39.72	0.71
Jarka	6.25	—
Karma	45.35	1.16
Kerwa	46.82	2.35
Kolabera	95.23	4.76
Parbat	40.82	4.05
Tilai	62.27	—

Source : Forest Survey of India, Purulia.

The picture is also same in other forest villages. It has become clear that unauthorised cutting of forests now forms one of the major occupation of the people of the most forest villages. A considerable percentage of population are fully engaged in wood cutting.

Pattern of Consumption

Wood is consumed for various purposes in the form of fuel, agricultural implements, constructional works as well as raw materials for wood-based industries. So the forests play a significant role in local and regional economy.

Fuel-wood Consumption

The per capita consumption of fuel-wood varies from family to family. About 90 per cent of the fuelwood is consumed by the rural population. Local people generally use agricultural waste, dried leaves and woods from village trees. A small fraction of the wood is purchased from the market. The consumption ratio of fuelwood also depends on the source of supply (Table 3.7). When the source of supply is from private lands, the consumption is less than the government land.

TABLE 3.7 : Fuelwood Consumption (per capita) in the Plateau Districts

Source	Per capita consumption (m^3)	Average per capital consumption in rural areas (m^3)
Private land	0.29	0.32
Government Land	0.42	

The rate of consumption of fuel wood is higher when the urban areas are located close to the forests (Fig. 3.1). The per capital fuelwood consumption in villages located within a distance of 10 km from the forests is mearly double than that of villages away from the forests.

Household Consumption

Next to fuel, wood has its importance in household use. Household

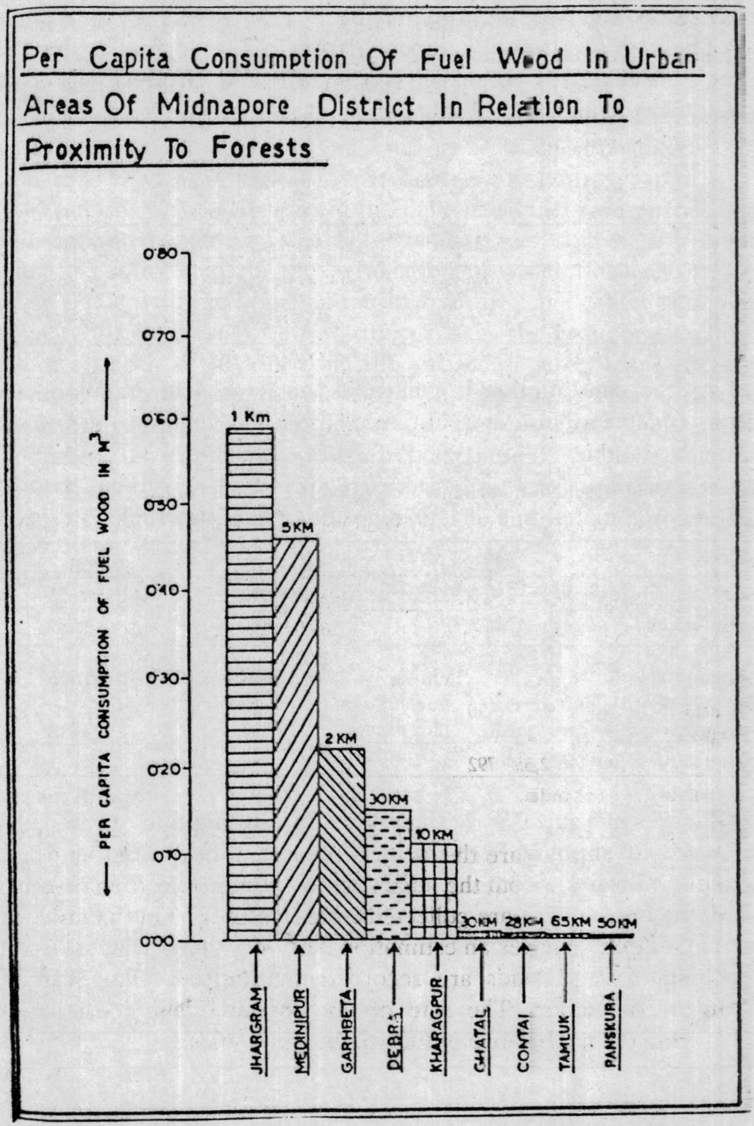

Fig. 3.1: Per Capita Consumption of Fuelwood in Urban Areas of Midnapare District in Relation to Proximity to Forests

consumption can be divided into three broad categories, (i) construction and repairing of houses, (ii) making agricultural implements, and (iii) furniture making.

These needs are usually met from adjacent forests by the villagers themselves. Bamboo is also a forest produce which is consumed at a high rate (Table 3.8).

TABLE 3.8 Total Annual Consumption of Bamboo

(*in tonnes*)

	Midnapur	Bankura	Purulia
Rural	194,335	70,194	53,985
Urban	4,018	5,799	5,336

As the consumption is double of the total yield, bamboos are imported from other States.

TABLE 3.9 : Household Consumption of Wood Per Year in Plateau Districts, 1983

Category	Midnapur	Bankura	Purulia
House construction	1,463,502	268,092	50,555
Furniture and fixtures	2,63,714	17,840	2,751
Agricultural implements	6,150	37,594	61,775
Fuelwood	2,658,792	712,200	1,197,918

Source : Forest Survey of India.

Figures in Table 3.9 are the estimated consumption of wood. All these are not supplied from the legal sources. It has been found that agricultural waste, which are collected by the villagers. On the other hand, the field survey gives an estimation that only 30 per cent of the cart loads and head loads are recorded. So there is mass scale poaching of vegetation. The rate of unauthorised felling can be assumed from the net balance of fuel, timber and poles.

TABLE 3.10 : Demand and Supply of Wood in Plateau District, 1983

			Net Balance of Fuelwood (in cu.m)
Distirct	Gross demand	Availability	New balance
Midnapur	2,659,792	1,092,180	(-)1,567,612
Bankura	712,200	81,439	(-)630,751
Purulia	1,197,913	300,835	(-)897,073
	Net Balance of Timber		
Midnapur	191,412	191,412	—
Bankura	76,731	58,429	(-)18,301.59
Purulia	N.A.	N.A.	
	Net Balance of Poles and Pulpwood		
Midnapur	112,912	(Poles) 28,942	(-)83,970
Bankura	14,200 (Pulpwood)	11682	(-)2,517
Purulia	N.A.	N.A.	

Source : Forest Survey of India, Government of India, 1984.

Thus there is a regular movement of forest products both from outside as well as within the district. A survey conducted by the FSI on the movement of forest produce outside the district and State indicate that mainly timber, poles, firewood, Biri leaves, Sal leaves etc. are sent outside from Purulia district. Timber, pole, firewood and pulpwood are sent from Bankura. No such data are available for Midnapur.

Role of Forests in Regional Economy

The forest play an important role on the regional economy. The distribution of wood-based industries, their transportation from producing areas to consuming centres exert great influence on regional economy. It opens large avenues of employment.

Forest-based Industries

No large scale forest based industry has been established here. The industrial activities in this region are still of artisan type. Amongst the industries, saw mills are most important. Supply of Pole and pulpwood to the mines and paper industries respectively is also very important. Other industries are culture of lac, tassar, silk weaving, mask making etc.

Consumption of Forests in Large-scale Industries

Supply of pulpwood and pole to the paper mills and collieries respectively from the forest department can be considered large scale enterprises (Table 3.11). Timber production is very low owing to the scarcity of suitable trees. The pulpwood is mainly supplied from plantation areas. Titagarh Paper Mill and Bengal Paper Mill are leading paper mills in this region and specific areas for plantation are mainly allotted to them.

TABLE 3.11 : Supply of Raw Materials to Pater Mills from Government Forests (Plateau districts)*

Year	Pulpwood	
	Quantity in ADMT**	As % of requirement
1980-81	35,600	31.0
1981-82	54,824	47.7
1982-83	42,116	36.6
1983-84	27,680	23.8
1984-85	32,144	28.0
1985-86	22,473	19.5
1986-87	14,892	12.9
1987-88	–	–
1988-89	1,654	1.4
1989-90	15,000	13.9

* Less consumption due to closure and lock out of the mills.

* Air Dry Metrict Tonne

Source : Forest Statistics of W.B. Forest Div., Government of West Bengal, 1989-90.

It has been found that the paper mills worked out most of the areas allotted to them, but it is much less in comparison to their requirement.

The average supply from the State Forest Department to the industry remains much below the requirements. This leaves a big gap between demand and supply at the State level. In this regards development of forestry is urgently needed.

The Forest Development Corporation is now supplying 300,000 sal poles and pit-props to the Eastern Coal Fields Ltd., from this region.

Consumption by Small Scale Industries

The saw mills are the only wood based industry in this region. Their total number and their estimated production is shown in Table 3.12.

TABLE 3.12 : No. of Saw Mills and Estimated Production

	Midnapur	Bankura	Purulia
Number of saw mills	48	20	28
Production per year	21,800	11,836	6,646

Except these saw mills, there are 2 to 3 units for building the bodies of trucks at Midnapur and Kharagpur which consume about 180 m^3 wood annually.

The important timbers processed are peasal, neem, gamer, sal, mahua, bahera, sirish, kend, arjun, chatian, jam, kathal, mangoe, neem, badam etc.

In Midnapur, saw mills are located all over the district. But concentration is more in areas where railhead facilities are available. There are about 48 saw mills in Midnapur. The total input of log is 27,266 m^3. According to the annual sawing capacity the mills can be grouped into several size classes (Table 3.13).

TABLE 3.13 : Size Classes of Saw Mills in Midnapur District, 1983

Size class mills by annual capacity	No. of Saw mills
Up to 200m^3	2
200-400m^3	17
400-600m^3	8
600-800m^3	10
800-1000m^3	8
Above 1000	3

Source : Forest Survey, Government of India 1984.

The saw mills with the annual capacity of 200-400m^3 are the highest in number. Most of them are located in the rural areas. These saw mills consume about 18.40 per cent of total timber. The major out-turn comes from the mills of 600-800 m³ size class. However, logs are also brought from outside the district, mainly from the Forest Development Corporation of Orissa. The important railheads where the saw mills are concentrated are Kharagpur, Midnapur, Balichak, Panskura, Chandrokona Road etc. In East Midnapur the source is mainly from private land and on the West from government lands.

It has been found that about 7.500 m³ or 27.50 per cent of total logs are obtained from private lands. About 18 to 20 or half of the total saw mills are fed exclusively by local timbers.

The units are mainly operated by hired persons because these are very much periodic in nature. Much of the work is done by hand. The machineries are old fashioned.

There are about 20 saw mills in Bankura district. The largest is the Sarenga Saw Mill with annual out-turn of 1.274 m³ and smallest one have the installed capacity of only 42 m³/annum. The average capacity of the saw mills is 500m³/annum.

Purulia district has 28 saw mills. Unlike lac and tassar, saw mills are the most important forest based industry of this district. At present the total annual production capacity is 15,045m³ but the total production is only 6,646m³ in 1983. This is much below their rated capacity. This is because of the insufficient and irregular supply of electricity and raw materials. According to their annual capacity the saw mills are grouped into following size class in Purulia district :

Size class by annual capacity (in m³)	No. of saw mills
300 or less	6
300-600	8
600-900	2
900 and above	2

About two-thirds of the saw mills possess annual capacity of less than 600 cubic metre. It has been found that about 50 per cent of saw mills of Purulia district import timber from Bihar, Orissa, Madhya Pradesh etc.

Thus it can be assumed that to run the saw mills, the largest forest based industry, supply of raw material is the prime necessity. Afforestation and rehabilitation of degraded forests are, therefore, necessary.

Cottage Industries

An insignificant percentage of population of Bankura, Midnapur and Purulia district are engaged in making woodenplie, combs and toys from timber like Papra (*Gardenialatifolia*) Gamar, Kurchi etc. *Mask making* from paper pulp is also a cottage industry in Purulia.

Infrastructural Development

The development of forest without adequate transport system is not possible. The forests are located in patches. So the exploitation and transportation of forest products both major and minor helps to develope transportation network which is mainly carried on by rail, road and river routes.

Employment

In this backward region there is no large scale forest based industry. for the absence of manufacturing or other tertiary activities, most of the people are non-workers or are obliged to work as marginal worker. Bulk of the workers belong to cultivators and agricultural labourer class. Growth of forestry and better utilization of forest products possess viability to provide extra employment to the people. Incidentally the forestry works so far remain manual i.e. seeding, nurseries, coppicing, felling, plantation etc. need manual labour. So forestry can provide significant avenues of employment. Now with the development of special schemes such as social forestry Drought-Prone Area Programme (DPAP) etc. the number of mandays in forestry are increasing steadily (Table 3.14).

TABLE 3.14 : Mandays Generated Under Development Schemes in Forests

Divisions	No. of mandays generated	% age of employment supplied
East Midnapur	408,477	–
West Midnapur	540,399	14
Bankura (North)	438,775	–
Bankura (South)	561,557	42
Purulia	247,337	13
Burdwan	279,456	6
Birbhum	205,793	10

Though the employment is not significant, still it holds promise for the future. It is important to note that total employment supplied by the forestry is much higher than employment in small scale factories.

Economic Out-turn

Inspite of rich mineral resources, the region continues to remain backward. No major industry has been set-up to boost up the economy. Though there are a number of lac, tassar and saw mill factories, these are not well developed due to lack of initiative of the people. Presently substantial amount of forest products, particularly from plantation areas, are exported to the paper mills of Raniganj, Calcutta, Hooghly and other places. So the possibility of augmemting employment in this sector appears to be bright.

The forest areas of the Plateau Section if properly managed, will play an important role in uplifting the economy of the local people. Better processing and manufacturing of the minor forest products are feasible through a network of small scale industries. This will absorb a good percentage of people. In this poverty-prone area, the success of better protection and management of forests rely much on our ability to create more employment and improving the standard of living of the people.

N.B. The micro-level survey in the plateau districts was done in 1983 by the Forest Survey, Govt. of India, and after that no such survey has been done.

4

ROLE OF FORESTS IN LOCAL ECONOMY: Village Studies

Introduction

Difference in physical and socio-economic conditions are the prime factors which cause diversities in the use of forest in this region. To assess the attitude of the villagers towards the forest, its use by them and to identify the present associated with the forestry, household survey has been carried out in five villages taking 10 per cent of the total households as samples. The villages have been a selected from Purulia and Midnapur districts located both close to natural and as well as plantation forest areas. One village is also selected from Midnapur district which is partly by dense natural vegetation and partly by plantation.

The villages selected for the case study are:

1. Jujardhara – An isolated tribal village in natural forest area of Midnapur district.
2. Ghatihuli – A tribal village in the natural forest area of Purulia.
3. Nepura – A non-tribal village in the plantation area in Midnapur district.
4. Ulidih – A tribal village in plantation area of Purulia district.
5. Arabari – A village with mixed forestry situations.

In the forest areas, the villagers largely depend on forests for their livelihood from early days, but nowadays to increasing deforestation, the area facing several hardships. In social forestry areas, the man made forests, solve part of their fuel problem but yet there are problems associated with the programme. The objectives of these case studies are to identify the problems of forestry so as to find out measures for their improvement.

Jujardhara village in Binpur Police Station of Midnapur district and Ghatihuli village of Bundwan police station in Purulia district are typical isolated tribal forest villages, where 70 per cent of the area is under forests. The villagers depend solely on the forests for domestic use as well as trade for a major part of the year except during the period of cultivation. But declining forest wealth is a matter of great concern bringing in its trail several economic and environmental problems. In this regards, regeneration of forest through afforestation programme is proposed. Plan to supply firewood from government depots are suggested. Arrangements of alternative income are also recommended.

Nepura village in Garbeta Police Station of Midnapur district and Ulidih of Purulia muffasil Police Station of Purulia district are typical examples of the villages where plantation under social forestry scheme has been found to be most successful. As such the barren lands of the area have now been transformed into green patches. Formerly these two villages had to face difficulties in the collection of fuelwood but now these have been substantially removed. The soil condition is also very poor to support agriculture which is the mainstay of their economy.

Though they have started plantation and have almost covered the major of the wasteland, yet it requires more attention. The problem lies in marketing.

Arabari village is an example where forest is being preserved by the tribals themselves and participating in plantation forestry also.

The present study brings out the multiple dimensions of the various uses of forest by the villagers. The success achieved in the social forestry has been studied. Suggestions have been made for further development.

Jujardhara

An isolated Tribal Village in Natural Forest Area of Midnapur

Name of the village : Jujardhara
Police Station : Binpur
Jurisdiction list No. 28

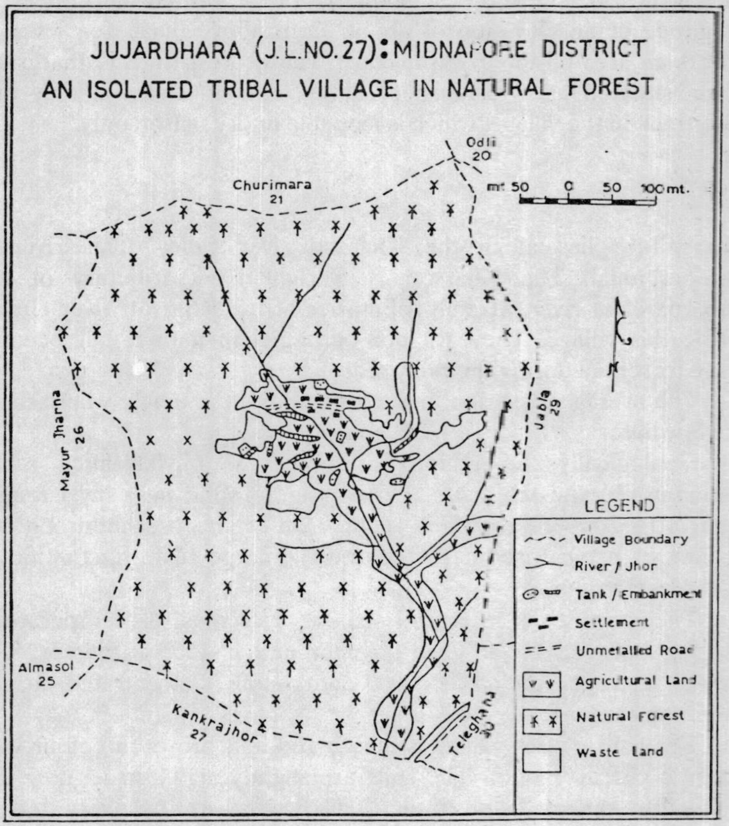

Fig. 4.1. : Village Map : Jujardhara

Location

Jujardhara (Fig. 4.1) is a small village completely separated from the villages by deep forests. The people of the village enjoys a primitive way of life. It is situated at about 22°43′ N latitude and 86°37′ E longitude at an elevation of about 225m above mean sea level. It covers an area of about 256 hectare. The nearest town is Jhargram, being situated at a distance of about 77 km. There is only one foot-track in the village, which is jeepable in dry season only.

Natural set-up

The village lies along the Kharsoti river valley. This river is non-perennial. The Kharsoti is a right bank tributary of the Subarnarekha river. Six sub-tributaries of the Kharsoti river almost girdle this village. They are non-perennial in nature, but become quite torrential during the rainy season.

The average elevation increases towards the north, where there are few hillocks.

Geologically the region is formed of phyllite-mica schist sillimanite biotite schist etc. The average groundwater level ranges from 8 to 20m. The level is around 2m in the post-monsoon but 15-20m in pre-monsoon. Thus there is a large scale fluctuation of water level in a year.

The climate is characterised by extremes of temperature, ranging between 20°-40°C in summer and 5°-15°C in winter. The annual average rainfall is 1,000 mm being concentrated in the monsoon season.

The soils of this village is mainly red and brown in colour with lateritic characteristics. The soils are slightly acidic in nature. The pH value ranges from 6 to 7. The soils are not very fertile. Jujardhara village is surrounded by dense forest, mainly formed of sal with mahua, asan, arjun, sisso, palash, neem, kend etc. as secondary growths.

Economic Base

Jujardhara is a small village with the total population of about 81 in 1971. The population was 61 in 1961, so there is about 33 per cent increase in population in a decade. The density of population is 0.3

person/ha. About 60 per cent of them are scheduled tribe dominated by Santal (Murmu, Tudu, Hazda).

The occupational structure reflects the backward economy of the people where 70 per cent of the population are non-workers. Amongst the workers, about 47 per cent are engaged in cultivation and others agricultural labourers. The male-female proportion of average working population is 2:1. The average size of a family is seven.

The house type reflects the typical character of this area. Due to availability of cheap and local wood, they use poles for house construction. Large size poles are used even for the farms of poultry birds. The walls are mainly of mud with rock slabs and thatched roofs. The homesteads are of compact type. Most of them live in join families.

Landuse

About 78 per cent of the total area is covered with forest and only 21 per cent is devoted to agriculture the balance represent homestead lands, roads and other areas. The farms are mostly rainfed as to provide support irrigation, in particularly good monsoons. Towards the south, the agricultural lands extend as strips on either sides of the Kharsoti river.

Amenities

There is acute shortage of drinking water. The three tanks in the village are dried up in dry season, which cause considerable hardship to the people. They have to collect water from miles away. The villagers construct small bunds in the low land (*Bahal*) and intermediate (*Kanali*) lands.

The village is an underdeveloped one. There is no post office. The nearest post office is at Kankrajhore, which is 3 kms. There is no medical facility. The nearest health centre is at Belpahari 30 km away from this village.

Relationship with Forest

About 78 per cent of the area is under forest and the village is girdled by dense forests. As such there exists a close relationship between the villagers and the forests. The dissected hilly terrain are characterised by rocky surface, which are not suitable for farming. The size of landholding is very small, ranging from 0.014 to 0.021 ha so the forest is mainstay of economy of the villagers. The mixed forest is formed of variety of species. The people of this region depend on the same to meet their domestic needs as well as materials for trade. Wood is the only fuel evailable to them. The demand of fuel varies according to the size of the family. About 15 kg. wood are needed per family per day. Previously only the twigs and leaves of the trees were used as fuels. But recently, scarcity of these materials have forced them to use stumps as well. The villagers often sell fuelwood to sustain themselves. The selling price of one bundle of firewood of about 20-30 kg weight is Rs. 9. Usually they sell it in the local market, but the same is sold from door to door as well. Even young boys and girls of 10 to 12 years of age are engaged in this occupation.

They also collect minor forest products which are found abundantly in the village. The minor products collected are sal seeds, mohua seeds and flower, kendu leaves, neem seeds etc. Besides there are other valuable plants like Myrobalam, *Bhang*, etc.

Among the forest species the sal (*Shorea robusta*) exerts much social and economic value in their daily life. They collect sal seeds and leaves. One kg of sal seeds costs 75 to 80 paise. One can collect 4 to 5 kg of sal seeds per day. The collections period of sal seed is very short. So in that period they are fully engaged in seed collection. The family on an average earns Rs. 50-100 worth of sal seeds in a year. This is not high in money value, but in this poverty prone areas, this brings some relief to the people. The women and children are usually engaged in this collection. Sal leaves collected from the forests are mainly used as plates. Plates are prepared by switching one leaf with another with thin sticks. One bunch of sal leaves with 1,000 leaves costs Rs. 10. Besides their use in the villages, leaf-plates are used widely in Midnapur, Bankura and Purulia town also.

Another important forest product, which is intimately connected with the social customs of the tribal people is Mohua fruits and flowers. Mohua flower is used to produce liquor which fetches high market value. Besides its domestic use, it is marketed in the neighbouring villages and towns. In years of drought, the mohua flowers are used as food. The fruits are used mainly for oil production. The seed oil is used in cooking, lightning etc. It is also used as massage oil.

Present Problems

The people of this village are closely associated with the forests. But now due to increase in population and unauthorised collection, the forest resources around the village are rapidly decreasing. As such, there is shortage of fuelwood. Dearth of mature trees has led to the dearth of sal and mohua seeds.

The farmlands of this village are without any facilities of irrigation. As such the farms are rainfed mono-cropped with low productivity (Table 4.1). Lack of infrastructural facilities such as absence of transport, market, post and telegraph office etc. have kept the people in a backward stage of economy.

Although about 50-75 per cent of the population are engaged in collection of sal seeds and kendu leaves, they are not getting remunerative price for these produces due to unscrupulous middlemen.

TABLE 4.1 : Landuse Pattern of Jujardhara

Landuse categories	Total area (Acres)	Forest (Acres)	Cultivated waste (Acres)	Cultivable waste (Acres)	Area not available (Acres)	Total
In acres	633	494	133	2	4	
Percentage of total		78.04	21.01	0.32	0.63	100

Proposals for Development

From the landuse Table 4.1 it can be assumed that forest has great impact on economy. So development planning for the village must include the forest also. The forest is decaying due to unauthorised exploitation by the traders and the local peoples. As such, the first step should be protection of forests by the local people for the local people. This can only be possible by making alternative arrangement of employment. With proper irrigation facilities, more areas can be brought under farming, thereby giving employment to more people. Shor bunds on the upper course of the river can store flood water for use in irrigation. The cheap fuelwood should be made available* to

* from the quick growing planted forest.

the people and arrangements should be make to supply coal. The minor forest products should be collected directly from the tribals. Though LAMPS (Large Scale Multipurpose Co-operative Society) have already established but it needs more initiative.

Afforestation through social forestry should be introduced which will supply the local needs. The tanrlands, hill slopes and adjoining fallow lands can be used for plantation of drought resisting species.

If the people could meet their primary needs from the afforested land, there is a possibility of its nuturing by the people as well. The degraded forests which have now been rendered unproductive, should be cleared off and afforestation should be made on a scientific manner under RDF (Rehabilitation of Degraded Forest) scheme.

Ghatihuli

A tribal village in natural forest area of Purulia District

Mouza – Ghatihuli
Police Station – Bundwan
Jurisdication list no. 89

Location

Ghatihuli (Fig 4.2) is a remote village with an isolated location at 22°50′N lat and 86°30′E long. This village is on the hilly uplands of Purulia with an average elevation of 300m above sea level. The village is situated on the side of a hillock. It covers an area of about 193 hectares. The nearest town is Purulia, located at a distance of 77km from it.

Natural Environment

Ghatihuli village is situated on the upper course of the Jora Nadi which is a distributary of the Subarnarekha river. The river passed through the village from north to south. The river is non-perennial. The average slope is towards the south. This village is encircled with dense forest of tropical deciduous type.

Due to the tropical location, extreme climate prevails here. The range of temperature is 20°-40°C but the maximum temperature often exceeds 45°C. In winter the minimum temperature come downs to 9°C. The average annual rainfall is 1,300-1,400mm. Rains are of short duration but torrential in nature. The relative humidity is very high and often reaches above 80 per cent.

The basic rocks are formed of phyllite mica-schist, sillimanite biotite schist etc. The soil is red and brown in colour. It is laterite in nature. The pH value ranges from 6.5 to 7. The average depth of groundwater level is 8-10m. In the post-monsoon months, the ground-water level is only 2-4m. below while it is 15 to 20m in pre-monsoon months. Thus there is large fluctuations of water level with considerable impact on the growth of natural vegetation.

The natural vegetations is of tropical deciduous type. Sal is the principal species. It is associated with Arjun, Bahera, Mohua, Teak, Neem etc.

Economic Base

Located in the midst of thick forest belt, Ghatihuli may be said to be an economically backward village. The recorded population of the village was 165 in (1971). This was 125 in 1961. The population density is very low, 0.85 persons/hectare. About 97 per cent of the total population belong to scheduled tribe. Only one per cent of the total population is literate. This is an indication of backwardness.

About 47 per cent of the total population are workers. Amongst them, 80 per cent are daily workers including agricultural labourer. For the scarcity of agricultural land, only 12 per cent are engaged in cultivation. More than half of the population are non-workers.

The people usually live in huts with thatched roofs and mudwall. Some of the hutments are sometimes surrounded by mudwalls. The logs are generally used for construction purpose. They are usually in place of Bamboo.

From the landuse pattern it is found that about 69 per cent of the total land is uner forests (Table 4.2). The area under agriculture cover is only 19 per cent of the total area. This limited area under agriculture is due to—

1. Paucity of Water
2. Undulating Terrain
3. Dearth of Irrigation Facility
4. Small Land Holding Pattern etc.,

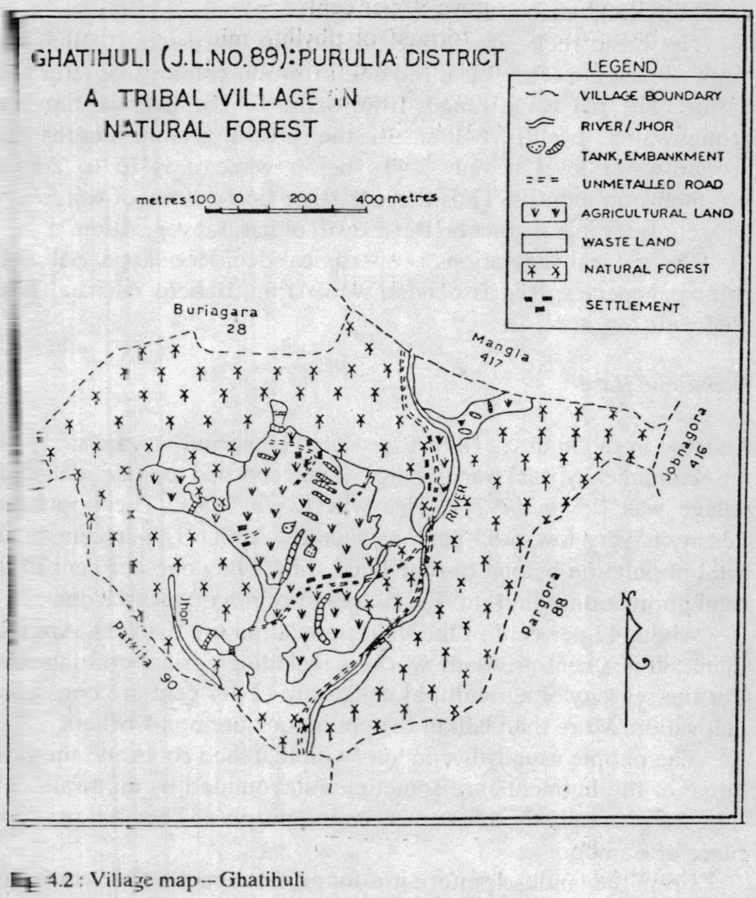

Fig. 4.2 : Village map—Ghatihuli

TABLE 4.2 : Landuse Pattern of Ghatihuli

Landuse Categories	Total Forest	Cultivated land (Unirrigated)	Cultivable waste	Area not available for cultivation	Total
In acres	332	90	22	33	477
In percentage to the total area	69	19	5	7	100

Due to the absence of irrigation, they have to depend fully on rain water. The rainfall is usually of torrential nature and of short duration. As such it is not always beneficial for agriculture. On the other hand, this excessive rain causes soil erosion enduring agricultural land unproductive. The percentage of cultivable waste land was five per cent in 1971 which has now been increased to the higher proportions mainly due to deforestation and soil waste. Along with it, about seven per cent of the total area is not available for cultivation. Some of the villages which are not under cultivation could be brought under agriculture with careful management techniques.

The source of drinking water is river, which is dried up in summer, when the villagers have to face acute scarcity of water. There are about 31 households in the village and no school.

The villagers adopt a technique of holding rain water by building small bunds along its course. On the flat topped land, agriculture is mainly practised, where water can be stored. The hill slopes or lands are mainly used for cultivation of inferior quality cereals i.e., Kodo, Gunja, Kurti etc. Maize is sometimes cultivated on their adjoining home lands.

Infrastructural Facilities

Being situated at a distance of 77 km from Purulia, the infrastructural facilities still remain poor. There is a single unmetalled road passing from north to south of the village. There are no market or post office facilities in this village.

Relationship with Forest

The villagers have close affinities with the forest. The undulating hilly terrain with tropical climate, encourages luxurious growth of vegetation of deciduous type. Sal as the principal species covering 90 to 95 per cent of the total forest. Miscellaneous trees such as Kend, Mohua, Sirish, Teak, Simul, Kusum, Arjun, Asan etc. are found as associates. The insufficient agricultural land does not ensure food for the whole year. So they have to depend on forest as the peoples of Jurandhara village.

Problems and Proposals

The villagers living in and adjoining the forest areas rely heavily on forest products for domestic use and trade. This result in considerable thinning or depletion of forest. Such unauthorised felling of trees have transformed the canopy forests into coppice one. Depletion of matured sal trees has resulted in a sharp decrease in seed production. On the other hand the coppice shoots of Sal plants are used to meet up fodder problem. Thus the chance of generating high forest in future is minimised.

For the high market demand, Kendu leaves are plucked indiscriminately which is injurious to the growth of species. The minor products are collected piece meal by the villagers. They do not get fair price of their produce due to lack of marketing facility and manipulation of middle-men.

Another vital problem is high incidence of unemployment. In the absence of any development programme, the villagers have to work as daily labourer.

Programme for the Future

The following programme may be taken up to tide over the problem :

1) The LAMPS which has already started functioning should be more organised. Intensive efforts for the collection of these products should be given particularly from the remote villages.
2) Arrangement of irrigation should be provided either through the construction of *Shor bunds* or from other sources. The water supply cold help to improve the productivity of foodgrains from the limited farm lands.

Increase in agriculture and subsidiary activities will indirectly lesson the pressure of population on the forest.

(3) Lastly, plantation through social forestry programme should be given due emphasis. In this dry areas, agricultural practice is less remunerative than forestry. If the *tanr* or hill slopes are used for plantation of eucalyptus hybrid, it brings more revenue and fuel to them. The villagers can use the leaves and twigs for fuel and sell the timber to pulp and paper production.

(4) Rehabilitation of degraded forest should also be encouraged. The areas with scrub jungles or Jhanti forest should be cleared off and new plantations should be made as quickly as possible.

Nepura

A Non-tribal Village in Non-forest Areas of Midnapur district

Police Station – Garbeta
Jurisdiction List No. 686

Location

Nepura (Fig. 4.3) village from Midnapur district is selected as an example where plantation under social forestry programme has reached to its optimum level. The location of the village is at 22°42′N latitude to 87°22′E longitude. The average elevation is about 100m. It is a small village with the total area of about 464 acres. Agriculture is the main occupation. It is well connected by unmetalled roads. The distance from Midnapur town is 39 km.

Natural Environment

This elongated village is bounded by the Kubai river on the north. The village is located within the Rarh Plain. The general height is within 75 to 150m. The Kubai river is a perennial river with a narrow course. Rocky out crops are found in the south.

This village is formed of older alluvium laterite. The soil is mainly formed of lateritic type. The pH value ranges between 6.0 to 6.5. The average annual rainfall is between 1,400 to 1,500mm. This

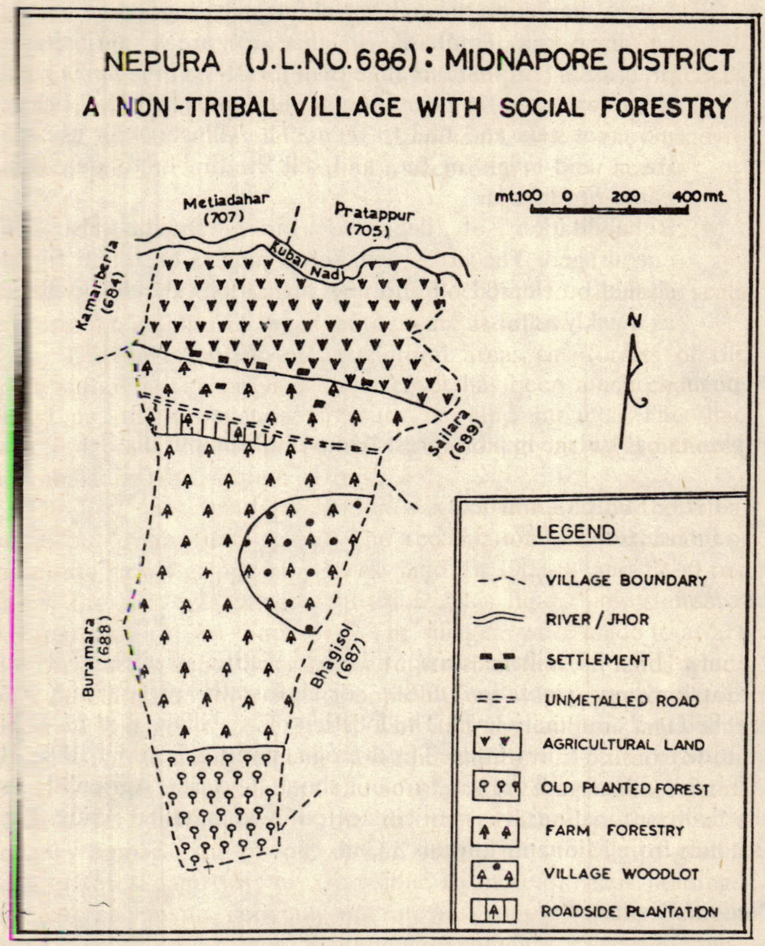

Fi_ 4.3 : Village Map—Nepura

region enjoys dry tropical climate. There is a great variation of temperature between summer and winter months. Though the average rainfall is high but it is concentrated during June to September. The other months remain dry. The average ground water level is 6m, which is 10-15m in pre-monsoon and 2-4m in post-monsoon months. Thus farming is highly disturbed by the fluctuation of water level.

Economic Base

According to the standard of the villages of West Bengal, Nepura is a moderately developed village. The total population is 459 in 1971 which was 332 in 1961 with occupied residential houses of 90. Thus there is a steady increase in population. The density of population is low with 0.98 persons/ha. Of the total population, about four per cent are scheduled caste and eight per cent scheduled tribes. The villages has one primary school.

The livelihood class reflects its poor economic condition. About one-fourth of the total population of the village are workers and the rest are non-workers. Amongst the working population, due to the scarcity of agricultural land, the percentage of cultivator is only 23 per cent while agricultural labourer forms the rest. The male and the females are working as daily labours. The average size of the family is 6 and the average working population per family is two. About 5 to 10 per cent of the people are government employees.

Most of the house have thatched roofs and mud walls. *Pucca* houses (cemented) with asbestos roofs are also found. Two or three rooms are attached together. Bamboo and interior quality poles are used for construction purposes. Some of them have domestic animal such as cow, goat etc. Poultry keeping is common, fetching secondary income.

About 11 per cent of the total area of the village are under irrigated agriculture. The farm lands are irrigated by tanks and canals. Canals are cut from the Kubai river. But the unirrigated land occupies 36 per cent of the total area, which is three times larger than the former. Due to the presence of highly degraded soil, about 23 per cent of the area remain as cultivable waste and 30 per cent of the area are not available for cultivation (Table 4.3). Thus the fallow land covers more than half of the land area of the village. There are few owners with large holdings. Rice is the main crop which is often cultivated twice in a year.

TABLE 4.3 : Landuse pattern of Nepura

Landuse category	Area (in acres)	Percentage to total area
Forest	–	–
Cultivated land (irrigated)	50	11
Cultivated land (unirrigated)	162	36
Cultivated waste	110	23
Area not available for cultivation	142	30
Total	464	100

The communication system of Nepura village is well developed. The roads are unmettalled but jeepable in all seasons. The village is well connected with other towns.

Relation with Forests

The village is almost devoid of forest. A small segment of the south of the village has old planted forest with mixed species. The common species found on the homestead lands are Peasal, Neem, Sirish and patches of bamboo. So there is a shortage of fuelwood. The villagers have to cover 2 to 3 km or even more to collect firewood. Besides shortage of fuel, absence of forest cover result in heavy soil erosion which restricts cultivation. With a view to check these problems, plantation under social forestry programme has been started in 1980. The plantation was started under the guidance of the school teacher, Sri Sadhan Chandra Pan. It was first planted on the school ground which was left barren for many years. Most of them are so interested that they are planting trees for many years. Most of them are so interested that they are planting trees in their homestead areas. People having 10 to 20 trees are not uncommon. Now *eucalyptus* has been planted all over the wastelands and in areas which are not available for cultivation. The plants with their high growth rate have form beautiful forest patch. Within 4 to 5 years, this village is likely to be turned into a green belt where trees would become large enough to protect the ground. Somewhere cashewnuts are also planted as ground vegetation. The participation rate is very high. This high participation rate in social forestry indicates their acceptance of this innovation. The villagers are usually interested in the plantation of eucalyptus. Today they have established a close rapport with the forests.

They are now collecting the leaves of Eucalyptus and Akashmoni from the ground for domestic use as well as for sale. It costs Rs. 5 per bag. This eases their fuel problem to a certain extent. The average demand of fuel per family is 10 to 12 kg leaves. They use the Akashmoni trees for making seesaw and crops in houses. The cashewnuts which have been planted as ground vegetation will become productive within the next few years and will become a remunerative crops.

Lastly it can be said the more than half of the land which was once left as barren, has now been transformed into productive fields.

Social forestry have covered almost all the waste lands. Amongst the four categories, farm forestry covers the highest area. Strip plantation is practised on the road sides but it is not evenly distributed. Village wood lot covers a small segment on the east.

Problems and Proposals

Though the villagers are interested enough in plantation due to tiny land holdings, they do not get sufficient return. The farmers are not sure of the market also. The problem of middle-men have already been started. As much for the future, the programme should be given broader dimension in terms of commercial viability. Though farm forestry covers a good percentage of area but large scale participation by the villagers (village wood lot) are almost absent, which needs more emphasis. The Government should assure them fair marketing facilities and restrict the activities of middle-men. Experiments and researches are necessary for the recovery of byproducts from the forest resources. From the administrative point of view, the programme requires more managerial personnel and better management. The panchayats can play an important role in this programme. To make the programme economically viable, it is necessary to grow other species having better economic value like planting of Sal, Arjun, Asan, Mahua etc. in this village.

Ulidih

A Community Forest Village in Purulia District

Police Station : Purulia Muffasil

Jurisdiction list No. : 224

Location

Ulidih (Fig. 4.4) village is selected because it is the place where community or social forestry has been first started in Purulia. The wastelands of this village have been almost covered up by this scheme. Ulidih may be cited as an example of successful social forestry village.

The village with an area of about 126.08 ha is situated at 23°10′N Latitude and 86°24′E longitude at an elevation of about 225m above mean sea level. Being situated in Purulia muffasil Police Station, it is well connected with Purulia (20 km) the district town, by metalled roads.

Natural Environment

Ulidih is located on the hilly uplands of Chhota Nagpur Plateau. The height ranges between 150 to 300m. The general slope is towards the north.

The climate is similar to dry tropical type. The annual range of temperature is about 20° to 40°C with highest temperature recorded in the month of May (48°C). The average annual rainfall is 1,300 – 1,400mm. Rainfall is mainly concentrated during July-September. Other months remain practically dry.

The Chaka river encircles this village on the west and south and finally towards the east. It is a non-perennial stream. The village is located on the western plateau and is mainly formed of granite and gneissic rock. The groundwater level fluctuated between 6-8m. The fluctuations are great between pre-monsoon (10-15m) and post-monsoon months (4m). The soils are of laterite type being mainly of sandy loam. The pH varies from 5.5 to 6.0.

Economic Base

Ulidih is a small village population of 224. The density of population is 1.7 person/ha in 1971. In fact, the population have decreased from 237 to 224 during 1961 to 1971 period. The village is dominated by both scheduled caste and scheduled tribe communities. There are only two literate persons in this village.

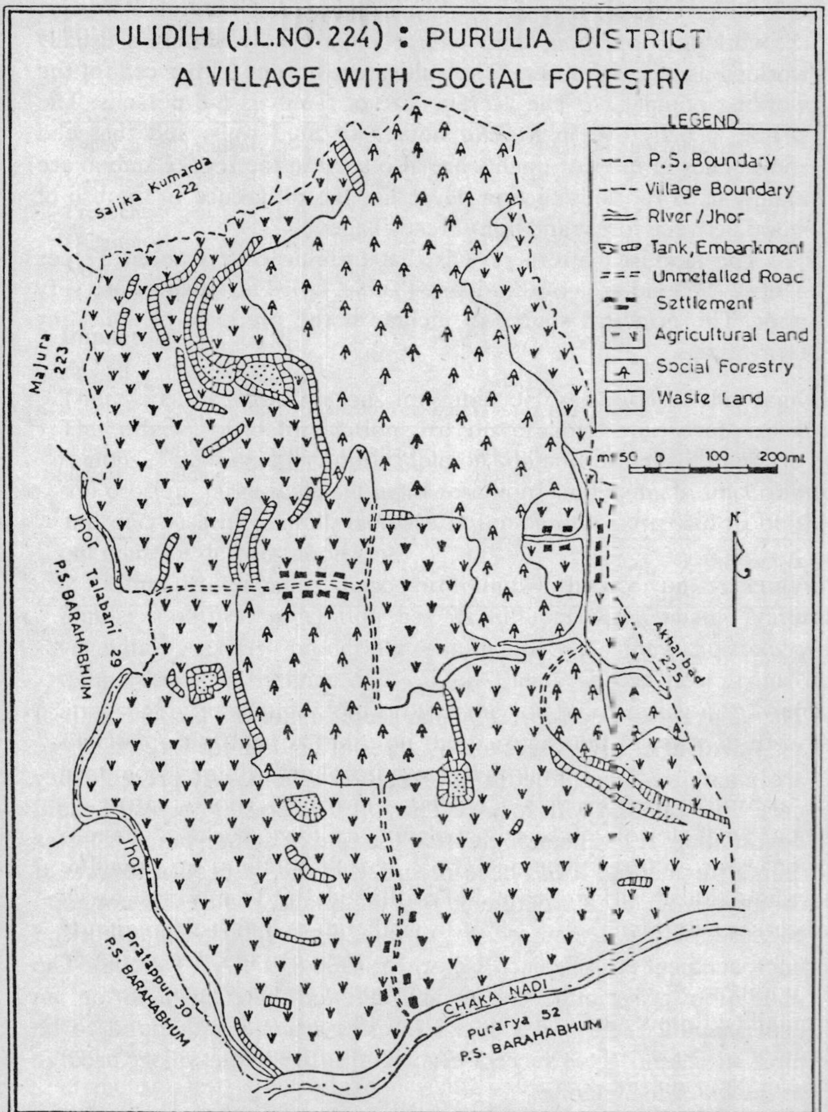

Fig. 4.4 : Village Map—Ulidih

About 55 per cent of the total population are workers and the rest are dependents. Thus the economic condition of the people of this village is better than those of Jujardhara and Ghatihuli. Amongst the working population, about 63 per cent are daily labourers, usually working as farm labourers. The cultivators occupy 37 per cent of the working population. The average size of family is 5-6 persons. The people usually live in *Kuccha* houses of mud walls and thatched roofs. Tiles of inferior quality are also used in the roof. Bamboo are mainly used for construction. Here lies the difference in the use of wood between forest and non-forest villages.

The landuse pattern reveals that farming covers about 37 per cent of the land area of the village (Table 4.4). The villagers are very poor. The principal source of income is the products of their tiny farm lands.

TABLE 4.4 : Landuse Pattern of Ulidih

Landuse	Categories	Total Forests	Cultivated Land		Cultivable waste	Area not available
			unirrigated)	irrigated		
Area (in acres)	327	30	100	30	6	161
%age of land	100	9	32	5	2	52

Relation with Forests

The village was a non-forest village according to the landuse map of 1954-57. But in 1971, there are 30 acres of land under forests. These are plantation forests under afforestation scheme. At present, they have little relation with forests. The forests also show signs of rapid degradation. It has been found that the rate of deforestation is faster than afforestation. It has been recorded in the successive reports of drought-prone area programme that the quality of forest is not good because of the uncontrolled grazing, illicit falling, unauthorised encroachment etc. (Project Report for 1980-81 DPAP, Purulia). The plantation programme was again initiated here in 1978 on an experimental scale under social forestry scheme and found to be most successful. Such success in social forestry in this village are due to the following factors:

(i) Presence of the tribal people in the peripheral parts of this village. The villagers are gradually transformed from farmers to marginal or agricultural labourers. The predominant monocrop and drought prone agriculture programme cannot provide them food for more than five months. Property is their life long partner. So through plantation they are trying to improve their economy;

(ii) The villagers have to move a long distance to secure firewood. Now they can use the leaves (*pala*) and branches of the trees after 2/3 years as fuel;

(iii) Moreover, the social forestry programme fetches good revenue by selling the timber within a short period. The cost of plantation is very low. Seeds or saplings and fertilizer are supplied free of cost from the forest department. Along with it, the remuneration of 10 paise in the first year and 14 paise in the second year is also provided by the forest department of each living plant. It has been found that as per the prescribed norms of spacing between the plants, 800 trees can be planted in an acre. Thus plantation of trees is found to be most profitable than any other activities; and

(iv) The species which are selected for the scheme are of quick growing varieties such as Eucalyptus, Akashmoni and a small percentage of Subabul. They take 6–8 years to become fully mature of selling and use as poles or pulps. The tops and leaves are used as fuel. the subabul itself is used as fodder to cattle and or as fish meal.

According to the villagers, before plantation the lands remained berren. Sometimes they were used to broadcast Kodo, Ganja or other inferior cereals. Due to scanty rain and poor soil, production is low. But the same lands have now been transformed into green with new plantations.

Problems

Though the common people have enthusiastically accepted the programme, the acceptance by the big and land-holders continues to be poor.

This shows that land-holdings under afforestation scheme are very small in size, which retards large scale production. So the initiative should be taken by large land-holders. In the absence of any fixed market, the villagers are suspicious about their profit. If there is no surety in marketing, the middle-men or traders will take the advantage and ultimately the area is likely to be grabbed by *mahajans* as in case of other items (Table 4.5).

TABLE 4.5 Size-class Distribution of Afforested Land at Ulidih

Size of the land (in acres)	No. of families interogated
Below 0.50	27
0.50–2.50	26
2.50–4.00	1
4.00–5.50	–
above 6.50	1

Suggestions

In this drought prone area, the poor villagers have the experience of being exploited and deceived from all rights of their own lands. They still have the idea that once their lands are afforested, it will be immediately vested with the Government. But social forestry is different from others. It is the only break-through amongst all other rural development programmes.

Though the programme has achieved some success, it should be remembered that it is not the task of an individual. Everybody should be aware of it. The Panchayats should be properly acquainted with the legal, extra-legal, economic and ecological aspects of afforestation programme and the role of social forestry in the revival of rural economy. The panchayats can convince the people at the village level about the real significance of this scheme.

The programme should be given more attention in terms of commercial viability. Plantation of different varieties of trees producing costly timbers or serving as lac or tassar host plants should be given preference. Mahua and Sal yield many by-products. The plantation of these trees will make the programme more lucrative. After all, one important aspect of the programme is to restore ecological balance through afforestation.

Efforts should be made to develop the system of organised selling of forest products either through panchayats or through the forest department.

Arabari

A Village with Mixed Forestry

Police Station : Garbeta
Jurisdiction list No. 452
District : Midnapur

General Outline

Arabari (Fig. 4.5) was a barren land after Second World War. Since then under the initiative of the Forest Department, the first plantation was started in 1949. The main species was Sal with associates of Asan, Sishoo, Teak, Kend, Mahua etc. By this process, the area as been gradually covered with dense vegetation. The forest are now considered as natural. Arabari village makes considerable use of the natural forest like Jujardhara village, but have also accepted social forestry programme. Under this programme almost all waste lands have now been covered with forest.

Location

The village is located at 22°42′N latitude and 87°14′E longitude with an area of about 630 acres. The elevation is about 30m. Arabari is the most important Silvicultural Experimental centre of the State Forest Department. Chandrakona Road, an important business centre, is located at a distance of only 15 km. Midnapur town lies at a distance of 31 km from this village.

Natural Environment

The Kubai river lies along the northern boundary of the village. It is a perennial river with very narrow course in dry season. The village lies at an elevation of 75 to 100m from the sea level, being located at the margin of the western plateau. The slope of the land is towards the west, rocky outcrops are also found in the villages

Geologically this region is formed of older alluvium with laterite rocks and soil. The pH value is about 6.0. The average ground-water level is only 6m. But there are large fluctuations between summer and winter months. In post-monsoon months, the goundwater level is

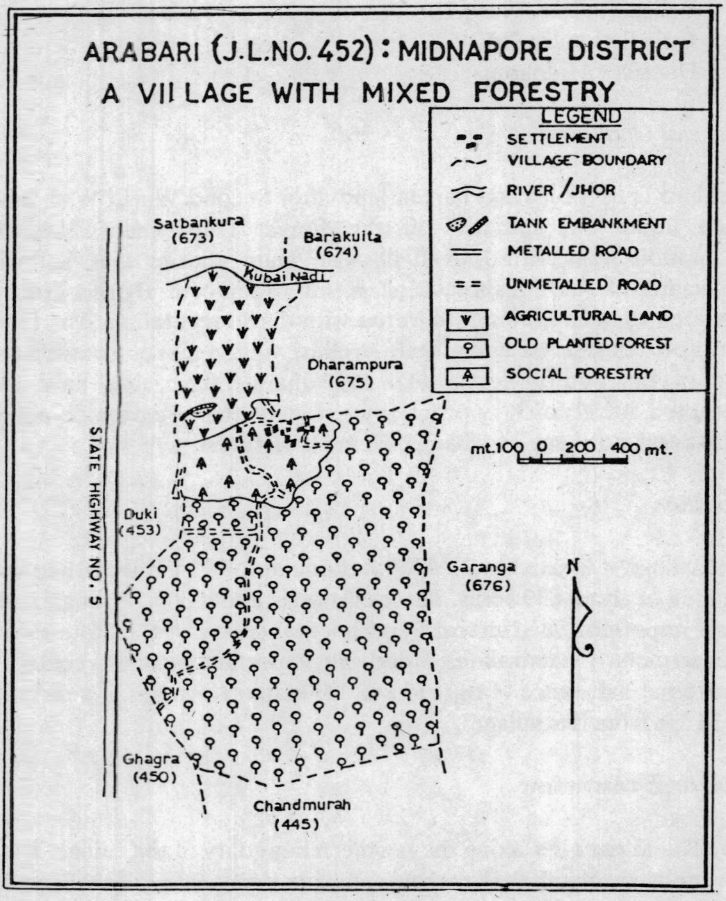

Fig. 4.5: Village Map – Arabari

2–4m. While in pre-monsoon season it is about 10–15m. This fluctuation retards the vegetation growth. With the high seasonal rainfall and temperature, the village enjoys an extreme type of climate. The average temperature ranges from 20° to 40°C. The highest temperature is recorded in the month of May. The average rainfall is 1,400 to 1,500 mm being concentrated during June to September. The relative humidity is very high and often reaches 80 per cent.

Economic Base

The village has a population of 198 in 1971. The figure was 126 in 1961. The population density is 0.9 per hectare. About 56 per cent of the population belong to scheduled tribe. The rest are scheduled caste, *bhumiz* etc.

About 28 per cent of the people of this village are workers and the rest are non-workers. Amongst the workers, only 14 per cent are cultivators. About 66 per cent of the working population are agricultural labourers, who are working against the daily wage rate of Rs. 5 to 7. About one-fifth of the people is engaged in forestry. Male, female and children are all working as daily labourers. The average size of a family is 5. The male-female ratio is 2:1. The people live in *kutcha* houses, being made of thatched roofs and mud walls. The boundary walls are often made up of mud. Drinking water is mainly collected from well.

At Arabari village, agriculture covers more than one-third (37 per cent) of the total area of the village. About 34 per cent of the village area is not available for cultivation. This shows that large segment of this village has been left unutilised. About one-fourth of the village area (26 per cent) is under forests which is mainly concentrated in the south (Table 4.6). The agricultural lands are

TABLE 4.6 : Landuse Pattern of Arabari village

Landuse categories	Area (in acres)	P.C. to total area
Forest	163	26
Cultivated land (unirrigated)	226	36
Cultivated land (irrigated)	6	1
Cultivable waste	20	3
Area not available for cultivation	215	34
Total	630	100

clustered in the northern part of the village due to proximity of the Kusai river. Irrigation facility is limited, covering only one per cent of the land. The size of the landholdings is very small, varying from 0.5 to 2 ha.

Rice is the main cereal. Vegetables and inferior quality cereals are cultivated on the adjoining homesteads. Double cropping is practised on irrigated areas. Elsewhere monoculture is usual. Most of the families possess livestocks such as cow, goat, buffalo, poultry birds etc. Sometimes, the livestock fetch side-incomes.

The village is well connected with neighbouring towns due to its location near the State Highway and railways. The State Highway No. 5 passes along the western boundary of the village. This highway is mainly used for the transportation of industrial raw materials. The settlements are of linear type, being set-up along the roads.

Relation with Forests

About one-fourth of the village (26%) is covered with forest. The principal species is sal. The forests play a vital role as it is found in natural forest area like Jujardhara, Ghatihuli etc. But besides the growth of natural forest, the social forestry scheme has made significant progress in the area. The villagers are mainly interested in Plantation because –

1. The tiny size of their landholdings is not congenial for large scale forestry.
2. Dearth of irrigation water does not encourage improvement in farm production.
3. Presence of vast waste-lands has made it easier to transform them into forest lands.
4. The highly degraded soil with rocky outcrop can only be used for tree plantation than other uses.
5. Use of the existing forest is not enough to supply the demands of the villagers. The principal species planted in this area is Eucalyptus. The leaves are now used for fuel.

Alongwith eucalyptus, cashewnuts have also been introduced. Some of the tribal villagers are marketing cashewnuts in this manner.

The success of social forestry at Arabari village is due to the following factors :

(i) The soil in the village helps in the growth of trees;
(ii) A good percentage of people in this village are land-less labourers. At present, the *patta* (free land) land has been given to them, thereby motivating them in the culture of forests; and
(iii) The people are innovation because of high literacy rate.

Arabari is a well developed forest village. About 20 per cent of the working population are engaged in forestry. Forest in this village is rather well protected. All the waste-lands have almost been covered up with planted forest.

Being benefited by the forest products, the people of this village are economically well off than their counterparts in other parts of the region.

Problems and Proposals

Though social forestry have been initiated at this village but individual participation is less than Nepura. It needs more attention. the area under irrigation is very low. Extension of irrigation facility by the government agencies may attract the people to agriculture. About 37 per cent of the total land are laying fallow which can be used in the production of inferior quality cereals as well as in plantation of trees. These measures will ease the pressure of population of the residual forest area of this village.

5

DEPLETION OF FOREST RESOURCES : A Temperal-cum-Spatial Analysis

General Outline

The forests are not more aggregation of trees. It is a biological unit having a vast social organisation of living communities at work. Trees play crucial role in conserving soil moisture. The quality of the environment depends to a great extent on the forest ecosystem which encompass the atmosphere, soil and its living organism. Forests also convert sunshine into various forms of energy as fuel, oil and oil products, which can be directly used by man. A majority of the population draw their requirements from forest for erecting their huts and other household implements. It is the home of wild life and promotes the beauty of nature.

But moderization does not pay much attention to the care of biotic communities. This is at the root of all the trouble. Destruction of forests and degradation of soil, water and air, are inherent in all modernising programmes. It has been stated that in our country indiscriminate destruction of forests takes place at an alarming rate because of socio-economic maladies. This can be visualized from the progressive decrease in the per capita forest area of the region. The per-capita forest area now stands at 0.02 ha (1981), as against 0.05 ha in 1951. The world average in 1981 is 1.08 ha. If this rapid rate of depletion of forest wealth is not checked, deforestation will reach to such a level, so as to threaten the very existence of mankind.

History

The history of forest in India is indeed one of deforestation (Chatterjee, 1964). It has been found that forests in general have suffered in the past through human agency and by the natural causes as volcanic action of great intensity, flood etc. but the share of man in the destruction of forest in undoubtedly the greatest. Man not only cuts down the trees to get building materials, fuel and medicine, but completely wipes out forests to acquire lands mostly for agricultural purposes. History shows that in India the ruthless destruction started with the settlement of the chalcolithic people in the Punjab, Sind and Gujarat in the period extending from 3500 to 3000 B.C. approximately. They came from the ancient kingdom of Elam, Akkad and Sumer in the Euphrates Tigris valley and wiped out the forests in favour of towns, villages, cultivation and pesture lands.

No firm date can be fixed indicating the period of deforestation. But in this plateau, available evidence shows that it was started towards the beginning of the 19th century, after the Permanent Settlement Act, when *jamindars* were faced with task of paying revenues to the State. Rev. A.E. Brown of Wesleyan Mission who spent 23 years in Bankura wrote that heavy cutting of sal trees did not start till 1914.

The first systematic inquiry about the condition of forests was made by E.O. Shebneare, the then Conservator of Forests in 1925. He dealt chiefly with Bankura where he found that the forests were cut to an almost incredible extent. In his Report he mentioned of the progressive thinning of forests along the edges of the blocks, exposed to villages and along the roads. The condition has not much changed today as was found by him about 60 years back.

The records of Birbhum also reveal interesting features of deforestation. The district today has about 160 sq. km of forest area and a lot of waste-land. But it was not the same a couple of centuries back. After the East India Company took over the *dewani*, there was a great famine in 1770 when villages were wiped out. So from official report of 1772 much of the cultivated land was then replaced by jungles. In 1780 it had already been turned into a dense forested area. But the history repeats itself. The picture found from the Gastrel Report in 1863 was quite different. At that period, due to unrestricted clearing of the forests, fertility of the soil was adversely affected. The areas which were once covered with thick jungles and grass have now been turned into sterile and barren waste. Again throughout the 18th century, deforestation was more encouraged when the forests were mainly used to produce charcoal.

In Purulia, forests have been under the management of private owners for long time. So in this period unscientific land use practices had led to unregulated failing of trees.

In Midnapur it was under the *Zamindari*, where they used to conserve their forests only to exploit it according to their needs.

Causes of Deforestation

Several reasons have been put forward to account for the large scale deforestation in this area. Some of them are as follows:

(i) Due to extension of agricultural areas the forests of the region suffered. The original forest cover has been much changed. With time, the population pressure on land mounts. This had resulted degradation of the residual vegetation, giving way to uneven and unbalanced distribution of forests.

The first phase of deforestation was not initiated by the tribal people but by the immigrants. The reclamation and extension of agriculture slowly gripped the forest land. In 1903-04 and 52.50 per cent of the area was under cultivation. This figure now stands at 64.30 per cent in 1981 (Table 5.1). The villagers were made to aware that forest areas could bring larger return than tenated lands which could be cultivated. Moreover, people do not want to go out from their own villages to be settled with the agricultural land at a distance. This in a way explain the rather indifferent attitude of the people towards acceptance of the surplus vested agriculture lands available for distribution in the districts. Naturally, there is a tendency amongst the people staying around the forests to grab a piece of forest land wherever possible, to increase their holdings. Such encroachments generally operate in groups. Purulia being a border district, record more illegal encroachments. Recent increase in the agricultural land is an indication of encroachment. Quite often such an increase has taken place at the expense of forest lands.

Districtwise increase in agricultural land is the indication of encroachment which takes place mainly on forest area. The reason is that the soil conditions of this area are very poor to support agriculture, except on the reclaimed forest lands.

TABLE 5.1 : Decadal Variation of Agricultural Land in West Bengal Plateau Region

(as percentage to total area)

District	1941	1951	1961	1973-74
Burdwan	67.32	68.17	67.05	69.52
Bankura	52.80	50.66	47.19	55.51
Birbhum	55.77	67.24	73.90	75.49
Midnapur	59.27	68.81	64.27	64.26
Purulia	–	–	33.83	44.05

Source : West Bengal, Statistical Abstract, 1976-77.

The increase in agricultural land is specially well marked in Birbhum, Purulia and Midnapur districts. Encroachment has taken place on the reserved forests as well as in other types of forests (Table 5.2).

TABLE 5.2 : Encroachment on Forest Area (1983)

(in ha)

Districts	Reserved forest	Other Forest	Total	% age in total
Midnapur	301.08	3480.20	3781.28	2.2
Bankura	–	3478.51	3478.51	2.3
Purulia	126.40	3541.60	3668.00	4.2
Birbhum	16.00	0.68	16.68	0.1
Burdwan	–	215.10	215.10	0.8

Source : West Bengal Forest Statistics, 1983.

Reports of the survey conducted by the Forest Survey of India 1983, have recorded the magnitude of encroachments on forest lands.

(ii) Subsequent clearing have been made for the outcome of large dams, irrigation canals, extension of roads and site for new industries and power plants. It is an example to show that about 3.6 per cent of the forest area was lost in the process (Tables 5.3 and 5.4).

TABLE 5.3 : Forest Area Lost in West Bengal (1951-81)

(in '000 ha.)

River Valley	Agri. purposes	Const. of roads	New industries	Misc. purposes	Total
1.90	313.70	2.60	2.90	3.92	325.02

Source : West Bengal Forest Statistics, 1983.

TABLE 5.4 : Forest Land Diverted/Likely to be Diverted for Non-forestry Purposes

Year	Forest Area ha	District	Purpose
1982	2.00	Midnapur	Construction of Eucalyptus Research Centre etc.,
	55.32	Bankura Midnapur	Kangsabati Irrigation Project.
1983	2.44	Bankura	Const. of 33 kv.O/H line.
1984	4.93	Bankura	
	4.58	Burdwan	Electric transmission line
	3.80	Burdwan	
	0.004	Midnapur	Contruction of repeater station by P&T Department.
1986	3.80	Burdwan	Laying transmission line.
	306.00	Bankura	Majhia Thermal power Project D.V.C.
1989	85.50	Birbhum	Construction of Bakreswar Thermal power station

Source : State Report on West Bengal Forests, 1988, 1988-89, 1990-91.

(iii) Growth of new towns and urban centres around industrial plants have caused further deforestation as most of the urban centres are stretching at the expense of forest areas. This is because the forest areas could keep down the financial cost of development. The Durgapur are in fact have been built by clearing of forests. The ever-growth demand for high quality wood for construction materials has meant dwindling of forests. Urban growth, even it is on the waste-lands, means destruction of forests. This is due to the fact that a large percentage of migrants in towns and cities continue to depend on wood. In their original villages they often use dried leaves and broken twigs as fuel. So the biotic communities are at the root of all this devastation.

(iv) Further deforestation took place with the opening of the railway line then known as Bengal-Nagpur Railway, connecting Kharagpur Jhargram (1898) and that of the branch line from Kharagpur to Adra through Midnapur and Bankura (1903). With these connections, the interior forests have become more accessible. Due to improvement of transport facility, the value of wood and forest products suddenly increased. Benskin (1939) in the Report commented that the opening of the East Indian Railway and considering the proximity of Bankura forest to Calcutta, the duration of forest will not exceed 50 years.

(v) The villagers of the plateau sections are very poor belonging mainly to scheduled castes and scheduled tribes. Absence of any large scale industry and dearth of alternative fuel, have made people rely on forest for their livelihood. The culture, life style and religious activities of the tribal are very much centred round the forests. Gradual depletion of forests causes socio-economic imbalance. Though the annual hunting ceremony is going on from age old days, the impact of it has become more pronounced in recent years with the increase in population and decrease in forest cover. The destruction of forests which takes place during this hunting ceremony has also become a matter of concern (Table 5.5).

TABLE 5.5 : Number of Tribals assembled and Animals hunted in Festivals in Purulia District

Year	No. of tribals assem-bled	No. of animals hunted					
		Barking dear	Wild boar	Pea fowl	Jungle fowl	Rabbit	Total
1982	20,000	3	9	3	20	23	58
1983	42,850	6	15	2	10	2	35

Source : Records of D.F.O. Purulia.

(vi) During 1800, all these forests were controlled by the 'Jungle Mahals' which were subjected to British rule through the East India Company. At that time, there was a great clamour for increasing the revenue. So the zaminders and tenure holders were compelled to increase the revenue by clearing the forests and bringing them under cultivation. At that time, deforestation proceeded at such a rapid pace that by the end of the 18th Century the ill effects of deforestation were in evidence.

(vii) Over grazing is the next important problem which results in rapid deterioration of forest area and hinder its growth. During grazing, the cattle damage young saplings. The tramplings of cattle make the soil compact. The pores of the soil become blocked and discourage plant growth. Again the pasture land is quite insufficient for them because there is little grass except in rains. Out of the total land area of the country, only 3-5 per cent are grazing lands. About 70 per cent of the forage needs of livestock are met from grazing lands, and the shortage is usually met from forest areas. The livestock population of this plateau is also high.

But compared to the number of livestocks the area under permanent pasture is indeed insignificant (Table 5.6).

TABLE 5.6 : Number of Livestock in Permanent Pasture

Districts	Livestock in '000	Livestock per ha in '000
Midnapur	2700	15.92
Bankura	1600	33.13
Purulia	1300	12.31
Birbhum	1300	8.46
Burdwan	1700	5.29

Source : Statistical Abstract, West Bengal.

So whatever pasture lands are available are subjected to over grazing. Over 40 per cent of the land are over grazed, 33 per cent medium grazed. Thus over half of the area is under heavy grazing. For this high rate of grazing, the natural growth of vegetation is restricted at the grass root level. Due to heavy grazing the sal vegetation is much damaged. Again, the cattle feed more on sal leaves than on grasses. In this way, sal forests at their peripheral areas are much damaged. March to July are the dry period when pastures become scarce, but it is the time when sal trees generate new shoots. These tender shoots become the first victims of the cattle. When this happens repeatedly, the saplings fail to grow and eventually die.

(viii) Forest fire is another cause of the rapid depletion of forests. In the plateau section, natural fire often takes place. Burning of forests, by the cowboys as well as by man to prepare land for agriculture are common (Table 5.7). Here, forests are burnt, so that ash may be washed down from it to the neighbouring fields. In fact channels are made to facilitate movement of silt and ashes to farm lands. This cause heavy destruction of forests.

During summer when dead leaves and under-growths are dry, such fires not only damage the young trees but also destroy the protective vegetative cover which act as a cushion of excess moisture thereby retaining water in the soil. In areas where there are dearth of firewoods or dry leaves, villagers lit fire in the forest in order to kill a portion of the coppice shoots which they subsequently claim as dead wood. As such they can collect the same for fuel. Burning of plants either in the forest or elsewhere ultimately lead to impoverishment of land, thus diminishing the fertility of the soil. On the other hand, it upsets the natural balance to a very marked extent.

TABLE 5.7 : Percentage Occurrence of Fire in Plateau Region

Types of fire incident	Per cent of occurrence
Very heavy	1.8
Frequent	13.3
Occasional	30.4
No fire	54.5

Impact

There are the basic reasons for rapid deforestation of this region. The effects of deforestation are discernable everywhere in the regions. The impact is appreciable on soil and drainage. Increase in soil erosion, large scale silting of reservoirs and cannals, uncertainty and decrease in rainfall, increase in temperature are said to be the out come of deforestation.

Climatic elements have profound influence on vegetation. Similarly natural vegetation has its influence in climate. Natural vegetation exerts moderating influence on the extremes of temperature in summer and winter. Plants absorb heat from the atmosphere in summer and protect the soil by releasing it in winter. Fluctuation of heat is more on the bare ground because it absorbs heat more rapidly than forest covered ground.

After a long controversy on the possible effect of forests on rainfall, it has now been concluded that the forests have some influence on rainfall (Warren, 1941). German foresters have shown in an experiment in the afforested area on the heath lands of Humburg that, due to the effect of forest, precipitation increases by about six per cent in a year.

Soil and vegetation together act as an entity. Vegetation is generally rooted in the soil from which they gain both mechanical and chemical support. Again when it dies, it releases chemical elements back to the soil. So a cycling pattern of nutrients take place. Thus the soil is the mother of vegetation. The soil erosion results from the misuse of the surface covering of the earth, whether it is due to destruction of forest, or by the misuse of arable or pasture land or by over grazing (Table 5.8). It has been stated that destruction of forests can do more damage in a region than a war. It could convert our land to a desert. Deforestation may act as a destablizing force in disturbing excosystem. The extent of damage caused by erosion is more on the grazed area than in grass and scrub lands. Erosion in this plateau area is considerable due to concentration of rainfall in a short spell of 4-5 months. About 500,000 hectares of waste-lands of West Bengal occurs in western lateritic tract which are subject to continuous erosion. This may be considered as a natural phenomena created by human interferences.

Soil erosion may be said to be an indirect outcome of forest management. Exposure of soil to weather depends on the methods of forest management. If the rotations are very short, the soil is exposed frequently to weather hazards and the chances of damage are more. In case of longer rotation the natural process of weathering and decomposition of humus and leaf litter replace the soil. The erosion also depends on the kind of soil cover.

TABLE 5.8 : Loss of Precipitation and Soil under Different Vegetation Cover

Forest cover	Loss of precipitation proportion of loss from normal	Annual loss of soil in proportion of acre
Forest with normal ground cover	1	1
Forest with poor grass cover	3	20
Well managed pasture	3	14
Grassland	10	130
Completely bare ground	25	3250
Agricultural crops	25	3250

Source : *Forest Ecology*, G.S. Puri, Vol.II.

So the soil erosion is more where the agricultural land and bare ground are abundant.

Drainage also suffers from excessive deforestation. With increase in population and extension of cultivated areas man as well as domestic animals gradually encroach towards the forest clad hill slopes. On the banks of the rivers, etc. The primeval forests which once covered the country are making their last stand in the hills and mountain systems. Not only the forests are indiscriminately destroyed, the undergrowth are also damaged by excessive grazing.

These are subject to erosion under heavy monsoon rain. The result is that thousands of tonnes of fertile soils are wasted into the rivers, thereby forming semi arid, in hospitable ravines. This results in an increase in the intensity of floods and reduce the subsoil storage.

In the earlier days the hilly catchment areas were covered with thick vegetation as in case of the Damodar, Ajoy, Mayurakshi, Kasai rivers, which retards run off and successively controls the intensity for floods. Now extensive deforestation is going on in the catchment areas as well. The silt deposition on the river beds is gradually increasing, thereby checking their beds and creating floods. The devastating flood of 1978 in the whole of south Bengal bears testimony of the effect. The sediment discharge of Panchet river is 10.00 ha. m/100 sq. km. The corresponding figures for the Mayurakshi, Kangsabati and Barakar rivers are 16.43, 9.65 and 13.10 ha/m respectively.

Alongwith the geological form, destruction of forests, unrestricted cultivation of wet paddy have resulted in soil erosion of a high magnitude. The extent of the problem in the Ajoy catchment can be judged from the fact that, out of the total area of about 250,000 ha in its catchment, about 72,000 ha. or one-fourth of the total area are susceptible to high erosion. Thus due to absence of forest cover, disturbance in the stream flow is recorded.

Hydrologic data from June to September when the forest is in full leaf, shows that while precipitation ranges from 32 to 50 cm., the average stream flow is only 5 cm/year (Kittredge, 1948; Column, 1955 and Hibber, 1967). This is because during the growing season evapo-transpiration absorbes the storage space. Deforestation and suppression of regrowth greatly modify the nature's capacity to regulate the flow of stream. Pierce *et al.*, (1970) have noticed that stream-flow generally increases at an average rate of 28.6 cm/year after cutting. Deforestation has strong seasonal effect (Hormbeck, *et al.*, 1970). During summer, stream flow usually records a dramatic increase after deforestation (Table 5.9).

TABLE 5.9 : Effects of Deforestation on Sessonal Stream Flow

Season	1966-67	1967-68	1968-69
June to October	332	248	164
November to May	4	15	72

Source : *Pattern and Process in a Forested Ecosystem, Borman, Likens*, 1979.

In the deforested watershed, soils become relatively dry because of excessive evapo-transpiration. Such soil has insignificant moisture holding capacity. The difference in storage of available soil water between forested and deforested area is 50–75 m. The summer rain on the already wet soils of the deforested watershed quickly generates large stream flow. It has been found during the mid-summer storm (29-30 July) of 1969 that precipitation of 137 mm generated an instantaneous peak flow of 305 mm/day on the deforested watershed as compared to 100 mm/day for the forested watershed (Hornbeck, 1973). Peak discharges are also common and more pronounced during summer in deforested areas. Along with the increase in stream flow deforestation eliminates nutrient intake and results in a whole array of changes which have detrimental effect on soil and stream flow. Thus there is a reciprocal relationship between drainage and vegetation. So loss of one will cause destruction of the other.

The experiment about the relation of runoff and vegetation (Ghosh, 1979) in Beliatore (Midnapur) has shown that—

(i) Run-off and soil loss is least under forest,
(ii) Run-off and soil loss is maximum in waste-land.
(iii) Run-off and soil loss is conspicuous in areas covered with bushy sal coppice, burnt artificially twice during the year.

It has been seen in an experiment at Jhargram (Ray, 1971) that areas having tall coppice trees and contour trenches of about 10 m apart, have only 1.5 per cent run off and it is 13.37 per cent in waste-land, where the plot is devoid of any vegetation.

The extrapolation of the results obtained from experiment come to a conclusion that with the destruction of these trees (forests) we are inviting more problems. However (1962) describes the role of forest in reducing flood flow in the following ways:

1. with holding water from the soil (interception).
2. protecting or increasing infiltration capacity.
3. reducing erosion and thus reducing the bulk of the flood flow by sediment.
4. maintaining and improving the soil permeability and large pore space within the soil to make more storage space and increase its effectiveness.

5. more rapid removal of water from soil storage between storms.

Experimental evidence indicate that interception by canopy and litter reduce surface run-off. These are because, in the barren land the raindrops strike the ground directly and tend to open the soil pores. The larger openings into the body of the soil are often choked with soil strained from muddy infiltration and make the soil loss absorptive which increase the rate of run-off and erosion. Gully erosion takes place thereafter. With little rainfall on the bare hills, small tributaries and sub-tributaries turn into torrents and rapids. These cause bank erosion on both sides of the rivers and bring the thick soil soup into the streams.

Suggestions can be put down to recover the drainage system :

1. First of all the detailed survey of land classification should be carried out. Each portion of the catchment should be delineated which will be permanently covered with forests.
2. The denuded areas on the catchment should be afforested and the surrounding forest areas should be kept as reserved forests.
3. Adequate provision for protection of forest area against fire, grazing, cultivation etc.
4. Conservation measure should be taken by planting and establishing of minor structures on the upper reaches of streams to prevent further erosion.

Along with the natural problems, deforestation will also affect the overall economy of the area. With continuous deforestation, there will be a time when there will be no further supply of fuel and timber available for use. This has been a reality at Taldangra-Sarenga village (in Bankura district) Sugnigerya and Sugnibasa village (in Midnapur district).

Deforestation also adversely affect the wild life. Forests act as the home of animals. From O'Malley's Report (1911) it is found that the carnivora of Midnapur were then represented by tiger, leopard, bear, hyaena, samber, spotted deer, etc. But now only small animals and bears are found. It should be noted that once the balance of nature in our forests is upset, we may have to spend lots of money for artificial protection of environment. Still they will not be the ideal home for them. Depletion of these animals will help in the multiplication of rats which will damage both the field crops as well as stored grains.

6

RECENT ATTEMPTS AT AFFORESTATION

General Outline

The forests of the plateau region of West Bengal show a complex life history. In the past, due to agro-economic movement all over the area, absence of any proper guidance, alongwith unplanned and uncontrolled expansion of agriculture, forest lands were cut off rapidly. Destruction of forests thus ensured quick cash return but the process brings in its trail disastrous effects on environment. In this process the original form and structure of the forests of this region have been extensively modified and indicate a very unbalanced spatial distribution. Large portion of the tract is covered with patches of forests interspersed with huge expenses of uncultivated lands or some of them are absolutely barren and tree less. This is because the forests were under private management for a long time. Random and reckless felling, unrestricted grazing, recurrance of annual fire etc. are some of the common features attributing to such degeneration of natural vegetation of the region.

Another contributory factor is the West Bengal Estate acquisition act. Though it was promulgated in 1953, the physical possession of most of the forests was taken by the end of 1955. This transitional period was marked by irregular felling at a such rapid rate. Both the owners and the public made desperate bids to get whatever they could get out of the forests before protective measures were introduced. From then, the problems of preservation and rehabilitation of lands in this semi-arid zone have been given attention, with a view to ensure better stability of the environment.

Another problem of the Region is the gradual erosion of the top soil due to faulty methods of cultivation. The surface soil is washed away in subsequent years as no preliminary bunds or ridges are made. The process is repeated till the underlying rock (morrum) is exposed. And then the area is abandoned. Thus innumerable barren spaces are found in this plateau areas. These are the areas which require immediate attention. In order to arrest soil erosion or to prevent the removal of top soils, better utilization of waste lands, for meeting local demand of fuel and timber, afforestation has been taken up on a large scale. Afforestation scheme forms an important competent and provide a viable alternative for growth and development in local situation. Though many other schemes have also been taken up for the restoration of environment, afforestation programme is probably the earliest scheme initiated by the State Government in this region. The programme of *Vana Mahotsava* or planting trees has been initiated since 1950 by the government to create awareness and enthusiasm in the minds of the local people for the preservation of forest and planting of new trees as "trees mean water, water means bread and bread is life". But the programme is yet to fulfil its ideas.

The Afforestation Programme has to take into consideration the following hazards:

(i) fight against drought and erratic rainfall;
(ii) ensure natural storage of under groundwater, including the revival of springs/streams;
(iii) Set up wind breakers by hedges thereby preventing erosion of soil by wind;
(iv) Making provisions of fruits, fodder, fuelwood, house construction material, medicinal plants etc.;
(v) working for purification of air by absorption of carbon dioxide, aerosols and metallic compounds; and
(vi) to make home for wild life

PLANTATION FORESTS

Apart from the plantation programme in the forestry sector, many other schemes are also concerned with plantation activities namely Drought Prone Area Programme (DPAP), Integrated Tribal Development Project (ITDP) special component plan for scheduled castes, Jhargram Development Project, Haldia Development and Digha Development Project.

Site selection for afforestation is indeed a difficult task. The primary consideration is the quality of land. Land which is not particularly suitable for other purpose is to be given preference for afforestation. The area should preferably be away from settlements. It is in the environs of settlements that the forests have to face various hazards and hence new forest areas should be preferably carried out in remote places for better returns.

Management

The afforestation programme in the Region was started in 1945-46. The work was first started in Midnapur District, viz at Jhargram Amlachati in (West Midnapur), Arabari, Lalgarh (East Midnapur) and Patrasayar (Bankura).

First plantation was started in 1946 with a conventional system, hoeing of land and making a ridge of worked soil, over which seeds were sown. But with the rains, seeds gradually washed away and there could be hardly any stock left. So in 1948 tractor ploughing was introduced and from then, the technique of countour trenching was introduced

Contour Trenches

The techniques which have been taken up are simple process of contour trenching and gully plugging. Different types of trenches and gullies were used. Main trenches are cattle proof trench all round the plantation. The contour trenches are 36′ to 75′ apart from each, 18" wide × 18' deep, all over the enclosed areas.

Selection of Species

The ultimate aim of afforestation is to bring the forest areas under reasonably stable formation, with the principal indigenous species of value within the shortest possible period. Initially a large number of species were tried in the plantations, but later emphasis has been put on sal, teak, peasal and lately on quick growing species like eucalyptus and Akashmoni. Amongst the species tried but not such favoured now are Minjire, Neem, Cashewnut, Siris etc. up to 1965 nearly 20,000 hectares of crops, including teak (700 ha) and of eucalyptus have been raised. But after that, quick growing eucalyptus has been preferred. This is with a view to create a protective cover of the soil, on which the principal species will be established. The quick growing species are mainly preferred because this tract has a great variation in the amount and distribution of rainfall. In some years, a long break in the middle of the monsoon causes heavy mortality of young seedlings. So importance are mainly given to the quick growing species, which could acclimatized the variations.

The species which are now widely used in afforestation are:

Eucalyptus hybrid
Akashmoni (*Accacia auriculiformis*)
Sal (*Shorea robusta*)
Teak (*Tectona grandis*)
Peasal (*Pterocarpus marsupium*)
Rosewood (*Dalbergia latifolia*)
Red Sanders (*Pterocarpus santalinus*)
Sissoo (*Dalbergia sissoo*)
Chatiwan (*Alstonia scholaris*)
Bahera (*Terminalia belorica*)
Asan (*Terminalia tomentosa*)
Arjun (*Terminalial arjun*)
Cashewnut (*Anacarduim occidental*)
Subabul (*Leucaena leucocephala*)

Of the grasses, Sabai grass is an important.

Many of the species are indigenous, only eucalyptus and Akashmoni are exotic to this area which are more important now because traditional forestry use to produce timber mainly.

Up to 1957-61, observation and experiments on sal, the principal indigenous species were carried out. But with the commencement of Third Five-Year Plan, programmes were taken for the plantation of trees of other economic uses as well. Thus eucalyptus and Kashmoni were planted in the region with a view to supply the raw material for pulp, chipboard and allied products to the paper industries.

Sal

Sal is not used as a major plantation species. Even though the highly degraded laterite soil in the region are able to maintain sal forest, technical problems, do not allow sal to regenerate artificially on waste-lands. Again the viability of sal seeds can be maintained only for a very short period. In most regions there is a time gap of over a month between sal seeding and onset of monsoon. Direct sowing cannot be done successfully. As the area suffers from chronic water deficiency, there is no question of irrigated plantation. 'Dona' planting can be practised. But when other suitable species with higher growth rates are available. There is no economic justification to take 'dona' planting of sal.

The growth rate of sal is very slow in comparison to eucalyptus (Table 6.1).

TABLE 6.1 Growth Rate of Sal and Eucalyptus

	Age (Years)	Length (m)	Girth (cm.)
Sal	5	1.5	1
	10	4	7
	14	8	2
Eucalyptus	5	6	7
	10	11	11
	15	17	30

Though the sal plantation have many disadvantages, still attempts are being made to grow it. Frequent failures of germination of seeds however retard its introduction in plantation area. In general it can be pointed out that the success of sal as well as afforestation, to a large extent become the gamble of the monsoons. Sal is a very sensitive tree. So plantation of it particularly provide risk. Sal is easily grazed. Again, over exploitation of bigger trees and processing of sal seed oil, leads to the scarcity of sal seeds. To cope up with all these hazards, eucalyptus and akashmoni are preferred as the main plantation species.

Eucalyptus

There are 600 varieties of Eucalyptus species with different morphological feature and physiological requirements. Among them *Eucalyptus robusta* and *Eucalyptus rostrata* are water consuming species which were planted in the marshy land in Italy and Kapurthala district of Punjab for reclamation. In West Bengal the species *Eucalyptus tereticornis* commonly known as Mysore hybrid is being raised. Small scale experimental plantation was started in 1961 and from 1962 regular plantations have been initiated. The reasons for acceptance are-

1. The most important quality of Eucalyptus is its versatility. It is an exotic species can be planted on a large scale under widely varied climate and edaphic conditions. The growth rate is 6m in 5 years whereas it is only 1.5m for sal.
2. On the field bunds and road sides it acts as wind break or shelter belt. Loss of organic matter due to complete decomposition is much less under it.
3. Owing to the oily texture of leaves and stem, only 11.56 per cent of total rainfall is lost by interception with the stand density of 1658 trees/ha and the balance is therefore available for absorption in the soil. In case of sal, such an interception is 25.3 per cent with stand density 1678 trees/ha.
4. As a perenial tree eucalyptus also restricts evaporation. The most important reason for its preference over to others is its role in the process of retrogression in soil. The soil condition reaches in such a degraded position for a long period of misuse and ill-treatment in this region that there is hardly any species except eucalyptus which can grow artificially or normally on them.
5. This species sends the tap root down to only 2m depth. Therefore, it is not possible for the plant to draw water from further down to deplete sub-soil aquifers. During the rainy season this tree draws the water through the rootlets and conserves this moisture in its body to last through the drought period.
6. This species requires much less water to produce 1 kg of wood as compared to any other indigenous species.
7. In Australia Eucalyptus has been extensively planted in catchment area of rivers for the purpose of soil and water conservation.
8. Eucalyptus species is never raised on lands where there are more than 200 viable stumps of sal. This is planted in barren wastelands or such covered with obnoxious shrubs and climbers, etc.
9. Eucalyptus is a multiple-use tree. It is used not only as pulpwood but for fuel, poles and posts and for making

charcoal. The mean annual increment of pulpwood is $5m^3$/hector/year ADMT and $7.5m^3$/ hector/year in green. Eucalyptus oil can be obtained by distilling the leaves. The fallen leaves of tree are being extensively used in South Bengal as fuel. Economic returns from the eucalyptus are so lucrative that an increasing number of people are going in for its plantation.

10. Depending on the quantum of sunlight reaching the forest floor there is either abundance or dearth of ground flora. If Eucalyptus is raised at proper wider spacing there will be plentiful growth of grass, climbers, etc. to meet the requirement of the villagers.
11. It has not been scientifically proved that pure Eucalyptus crop is bad for the environment. Eucalyptus plantations harbour birds, rodents and other animals.
12. Eucalyptus reproduces both sexually and asexually. Asexually it can be induced to propagate through root and stool cuttings in suitable condition by air layering. Instead of it, regeneration through seed is the best method. It is also a good coppicer from 10 years upto 22 years. The percentage of eucalyptus in plantation areas varies from 25 to 92 per cent in laterite zone.

Akashmoni

Akashmoni (*Accacia auriculiformis*) is the second important planted species covering 50 to 60 per cent of the planted area in West Bengal. Akashmoni is a native of Australia. It is growing artificially from 1946 in association with other species in laterite areas of West Bengal. The most outstanding feature of this species is that, it can grow on so poor soil where few other species can survive. Before the introduction of eucalyptus hybrid in 1962, it was the only resistant species. About 1,000mm to 1,400mm of monsoon rainfall with mean annual temperature of 26° to 32°C is favourable for its growth. Being a very adaptable plant, it can be grown on wide range of deep or shallow soils including sand dunes or on soils rich in mica schist, compacted clay, limestone, podsol, laterite and lateritic soil. The pH value of the soils may vary from 3.5 to 8.5. Several experiments on this species have revealed the following advantages of this plant:

(i) It is easy to grow and fast growing. The production rate is about 6^3m per ha per annum;
(ii) It is evergreen with thick leathery leaf and even in summer it provides effective shade and soil cover;
(iii) It is an excellent seed producer and regenerates freely even under the shades of other trees;
(iv) It can withstand root competition which is an important criterion needed for this locality;
(v) It has better growth than eucalyptus in clayey soils which is quite common in the plateau region;
(vi) The most important promise of Akashmoni is that it is a source of wood pulp; and
(vii) Due to its evergreen nature, it can be used as attractive ornamental tree withstands all conditions better than any deciduous tree.

Though it has got so many qualities, some disadvantages are also found during the experiment. Such as:

(i) It is less resistant to fire than eucalyptus;
(ii) Though it has good root system, it cannot withstand severe drought conditions, because the roots are closer to the soil surface;
(iii) Akashmoni has large lateral branches which break easily in strong winds;
(iv) It has a poor form for a timber species; and
(v) Up to one year, it is easily grazed. So it needs grazing control which is a problem in this region.

However, as the advantages of Akashmoni far outweigh the disadvantages, it has become the major crop in some areas and from 1973 onward it occupies 50 per cent of the plantation area in laterite tract. Now Orissa, Bihar, Uttar Pradesh, Karnataka are also using this species. In semi arid areas it acts as a good substitute for eucalyptus hybrid. The supply of N.P.K. helps it to grow in areas of poor soil.

Casuarina

In the coastal zones of Midnapur, Casuarina is the main planted species. The first nursery was established in 1945 at contain with the sole object of afforesting the sandy wastes all along the coast line starting from Khejuri in the east, right upto the Orissa border in the west. The first plantation was at Junpur and Digha in 1946. Gradually the planting series were extended up to Nijkasba, Khejuri in fairly extensive sandy wastes. In the beginning, Akashmoni and cashew nut sown in between the lines of casuarina later, casuarina was planted pure. The coastal plantations of casuarina have now created a green tree belt all along the foreshore of the Bay of Bengal. This tree belt also supplies firewood for local consumption.

Teak

In suitable soil, teak is also planted with wide spacing, keeping in view that it will not interfere with the growth of eucalyptus. It is grown as an insurance against any unforseen behaviour of eucalyptus in future. Atabada beat in East Midnapur division is an important area of Teak plantation.

From Table 6.3 it has been found that the plantation is carried on in all the districts steadily but the percentage to total geographical area is very small (Fig 6.1)

Urban Forestry

From 1976 onwards urban forestry has been taken up. Works are mainly carried on in the district headquarters. Recreation parks have been created and maintained by this urban forestry department.

Outlook

In conclusion, it can be said that afforestation programme can only be successful with government initiative alongwith the co-operation of the local people. Unless the people get adequate employment through various programmes of rural development, they will continue to depend on the forests to meet their immediate needs, though it is not sufficient to meet all of their requirements. With the success of afforestation, the planted species can make denuded bare hills their niche and gradually cover the rocks and create fresh and thick soil cover and supply the raw materials in the allied industries.

TABLE 6.2 : Area Under Plantation

	BP*	NA*	Total
Midnapur			
1986-87	32.19	64.46	96.65
1987-88	44.19	125.43	169.62
1988-89	28.52	92.25	120.77
1989-90	32.09	86.20	118.29
1990-91	41.13	83.55	124.68
Bankura			
1986-87	40-83	59.46	100.29
1987-88	30.31	43.98	74.29
1988-89	16.59	20.00	36.59
1989-90	25.40	39.60	65.00
1990-91	33.66	37.67	71.33
Birbhum			
1986-87	13.61	10.69	24.30
1987-88	12.00	34.25	46.25
1988-89	6.40	34.39	40.79
1989-90	4.00	25.60	29.60
90-91	12.50	24.00	36.50
Burdwan			
1986-87	11.64	9.22	20.86
1987-88	7.02	43.39	50.41
1988-89	3.93	36.50	40.43
1989-90	3.26	25.95	29.21
1990-91	9.02	46.41	55.43
Purulia			
1986-87	9.36	1.77	11.13
1987-88	15.52	53.94	73.46
1988-89	16.07	51.91	67.98
1989-90	17.16	47.60	64.76
1990-91	28.88	92.50	121.38

Source : Forest statistics of West Bengal Planning and Statistical cell, W.B.

B.P. = Block Plantation

NA = National Area

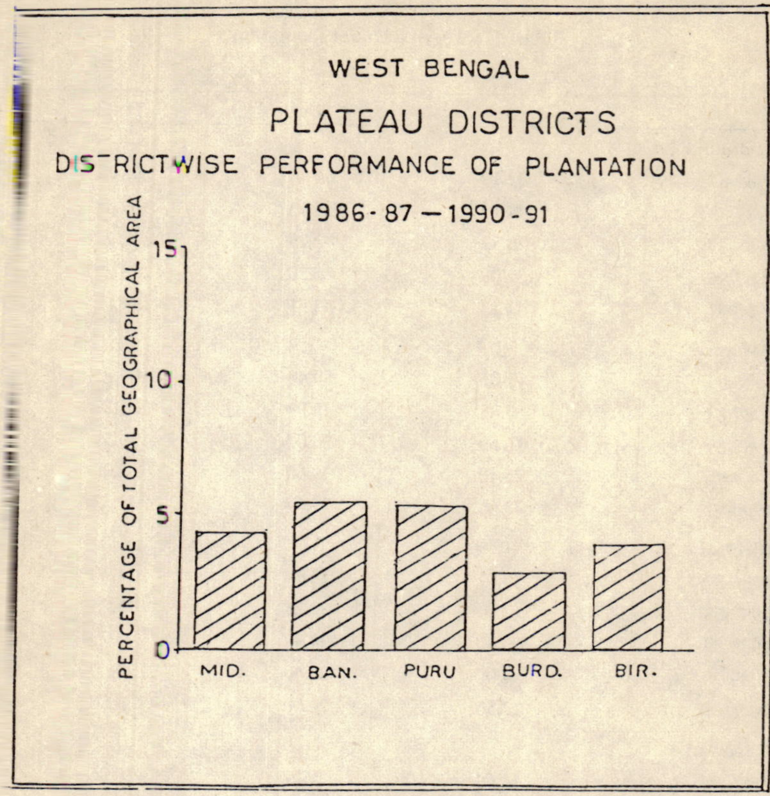

Fig. 6.1 : Plateau Districts : Performance of Plantation, 1986-87 — 1990-91

TABLE 6.3 : District-wise Performance of Plantation

District	Geographical area (in sq. km.)	Plantation area for 5 yrs. (in sq. km.)	Percentage to total area
Midnapur	14081	630.01	4.47
Bankura	6882	347.50	5.05
Purulia	6259	338.71	5.41
Burdwan	7024	196.34	2.80
Birbhum	4545	177.44	3.90

SOCIAL FORESTRY

Social forestry or community forestry is now being tried as an alternative measure which is done both by the private and government entrepreneurship. This adds a new dimension in the programme of regeneration of forests in this region. The concept of social forestry has been initiated by the National Commission on Agriculture (1976).

Objective

The objectives of social forestry are :

(i) to improve rural welfare by increasing the productivity of uncultivated lands;
(ii) to reduce destruction of forest resources by providing economic alternatives and unifies public and private interest; and
(iii) it also encourage people to regenerate forests for the purpose of which, necessary facilities are provided.

Conceptually the aim is to create conditions for the improvement of the rural areas and to develop awareness amongst the forest dwellers about value of forests. It is operated in the interest of rural communities residing in or near the forests. The programme is run as far as possible with the local residents. Operationally social forestry aims to combine the idle lands, labour and water resources to the best interest of the society in such a manner that the community becomes self-sufficient in respect of farm manure, fuelwood, fodder, small constructional timber, fruits, flowers, seeds etc. In this process, the soil will also be protected from erosion on a sustained basis. The most important indirect effect of social forestry is that it ensures the protection of local environment and ecology. Recreational and aesthetic amenities of the tree cover are the by-products of social forestry.

Relevance of the Scheme

A major portion of the south-western tract of West Bengal comes under the drought prone area of the State due to low and variable rainfall. Scarcity of water and soil erosion are common features in this region. The people of this region are poor due to low agricultural productivity and absence of other avenues of income, either in industries or in tertiary occupations. The land of this region are occasionally left fallow due to scarcity of water and degraded soils. As such most of the people living in the forest and its fringe areas have to depend fully on forests to maintain their livelihood. It has been estimated that about 36 per cent of the tribal people of West Bengal living in forests and its fringe areas, depend on forests to sustain themselves. The corresponding figure for the plateau districts has been estimated as 40 per cent. The highest dependency rate of tribal population is in Bankura and Midnapur districts (Table 6.4). The per family fuel requirement of this region has been calculated as 3.5 kg per day. In addition, small timbers are also used for the constructional purpose as houses, carts etc. Quite a number of families manage to earn their livelihoods by selling the wood to the villagers. All these problems are supplemented by production oriented social forestry programme. Again it has particular significance in this region for the regeneration of forest in the context of maintaining the ecological balance.

TABLE 6.4 : Percentage of Tribal Population Depending on Forests to Maintain their Livelihood

Districts	No. of Tribal population	Tribal population living in forests and its fringe areas	Per cent of people depending on forests
Bankura	208,735	125,241	60
Midnapur	442,963	262,729	59
Purulia	313,793	126,389	40
Burdwan	228,605	24,051	11
Birbhum	125,250	32,401	26

Source : Forest Directorate, West Bengal, 1983.

In social forestry the initial costs of seeds, manures etc. are supplied by the State. The planting of trees, their care and environmental needs are looked after by the local people. To protect the forests from the villagers as well as to support the socio-economic needs of the people, social forestry seems to be the most appropriate scheme. Its relevance lies in the protection of environment, production of fuel wood as well as to generate employment opportunities in the villages.

Environmental Requirements

About 33 per cent of land area should be under forest cover to maintain ecological balance. But in the plateau districts only 13 per cent of the land area are covered by forests. From the satellite imagery it has been found that only 10 per cent is actually under forests. Addition of social forestry will help to improve this figure. Another environmental advantage of social forestry is its role in preventing soil erosion, restoring soil fertility and in improving the micro-climate in the long run.

There are two aspects of this programme, which are specially emphasized viz., fuel wood production and employment opportunities.

Fuelwood

According to the Report of the National Energy Council about $0.28m^3$ fuel is needed per head annually which is equivalent to 0.2 tonne of fuel wood. On this basis the requirement of fuel wood for the different districts has been estimated (Table 6.5).

TABLE 6.5 : Requirements of fuelwood in the Plateau Districts of West Bengal

Districts	*Population in lakh	Demand of fuelwood (in '00,000 tonnes)
Midnapur	67.24	13.45
Bankura	23.74	4.75
Purulia	18.55	3.71
Burdwan	48.09	9.62
Birbhum	20.95	4.19

*1981 Census.

35.72 million tonnes of fuelwood is required for only five districts. Only 10 to 15 per cent of the total fuel requirements of the region are met by coal. So bulk of the need is met by fuel wood, agricultural residue, cowdung etc. The per capita annual requirement of fuel in the villages has been estimated as 0.38 tonne (in terms of coal replacement). The said figure for the town is 0.40 tonne. Supply of coal is much higher in towns than in the villages. So villagers have to depend fully on fuelwood. Only 10 to 15 per cent of this fuel need is supplied by Forest Department. Naturally the villagers have to collect the same from illegal sources. They also have to depend on agricultural waste and cowdung which could otherwise be used as fertilizer in the farm lands.

As per the assessment of the State Forest Department, the average annual production of fuelwood including government and individual, from all plantation would be approximately 105,000 tonnes after 10 to 15 years from now. This would be sufficient to meet the fuel needs of a major percentage of the people at the present rate of consumption. The annual increase in fuelwood production over 30 years (160,000 tonnes) would be equivalent to an annual saving of 400,000 tonnes of dung cake which could be utilized as organic manure to increase agricultural production. This would help to raise 25,000 tonnes of additional food grains in the plateau region of the State.

Employment Opportunities

The plateau section is a backward area where percentage of non-workers is more than workers. This is due to the lack of any important raw materials, dearth of industrial development and low agricultural productivity. So employment opportunities are limited (Table 6.6).

TABLE 6.6 : Working, Non-working Pupulation in the Plateau Districts of West Bengal

Districts	Workers	Non-workers
Midnapur	29.27	70.73
Bankura	29.06	70.74
Purulia	31.08	68.92
Burdwan	28.01	71.99
Birbhum	28.10	71.90
Hoogly	28.54	71.46

Source : Forest Statistics 1983, Census 1981.

It may be noted that more than two-thirds of the total population are non-workers which exert great pressure on the overall economy of the region. In this context, social forestry wing has a great role in creating additional employment opportunities. In total, some 15 million mandays were required for successful implementation of social forestry programme. This has provide full time employment of 9000 people. In practice, most of the forestry operations are usually concentrated over a period of 2-3 months of the year. This will be beneficial to the poor, since most of the annual jobs like digging pits and trenches are done by them.

Availability of Land

In the plateau districts a large percentage of area remains unproductive (Table 6.7). So there is no scarcity of lands for plantation. This unproductive lands are vested to the peoples which on the other hand encourage the social forestry programme (Table 6.8).

TABLE 6.7: Landuse : Plateau Districts

District	Area accordinig to village papers ('000 ha)	Area under forest (%)	Area not available for cultivation (%)	Other uncultiv-ated land excluding current fallows (%)	Current fallows (%)	Net area shown (%)
1	2	3	4	5	6	7
Burdwan	700.55	4.33	16.91	8.63	2.35	67.78
Birbhum	451.44	3.50	11.18	9.04	1.09	75.19
Bankura	685.58	20.61	10.92	13.36	3.03	52.08
Midnapore	1360.63	12.63	13.81	9.09	1.56	62.91
Purulia	623.41	14.05	14.75	19.61	5.26	46.33

Methods of Implementation

The plateau districts have the labour-surplus and capital scarce economy which bring in its trail the basic problems of poverty, low productivity, unemployment etc. Social forestry is one of the methods which can be extended deep in the rural areas and can trigger off a host of down stream activities. The social forestry was launched in 1981-82 which continues for 10 years within that period there are additions and alteration. The planting programme have been divided into four sub-component. Each plantations have differing concepts, establishment, management, has vesting, marketing, legal and funding features, which distinguish them from one another. The four categories are village wood lots strip plantation, farm forestry and Rehabilitation of Degraded forests.

TABLE 6.8 : Distribution of Surplus Land and Number of Families benefited (31st December 1980)

District	Agri. land vested ('000ha)	Non-agri.-land ('000 ha)	Total land vested ('000 ha)	Area of land distribution ('000 ha)		No. of beneficiaries ('000 families)			Additional No. of beneficiaries who can be covered ('000 families)
				SC	ST	SC	ST	Total	
Burdwan	31.6	10.9	67.6	35.0	14.2	38.2	17.3	87.4	330.3
Birbhum	15.6	16.2	39.5	20.4	8.3	21.9	9.5	42.9	205.6
Bankura	23.3	12.3	170.3	38.8	15.7	33.9	13.3	64.0	630.2
Midnapur	113.9	43.7	309.9	162.8	65.9	96.5	76.9	335.4	1243.1
Purulia	35.9	24.1	113.5	35.9	14.5	11.9	12.0	38.9	265.9

Source : Government of West Bengal, "Land Reforms in West Bengal : Statistical Report V", Statistical Cell, Board of Revenue, 1981.

Village Wood Lots

Under this category tree planting is done by Government or other institutions in collaboration with village people on community land. Cost of planting and maintenance is done by social forestry wing. The village panchayat used to protect and manage the plantation for two years in co-operation with social forestry wing.

Ulidih village of Purulia and Chandrapura village of Birbhum districts are such centres. In Birbhum Arjun trees are mainly planted which act as host-plants of tassar culture. Production of tassar has already been started thereby providing employment to the villagers.

The total area under village wood lot plantation is only 1390 ha but appraisal target was 4200 ha. So actual achievement is only 33 per cent. This is mainly due to the non-availability of Panchayat-owned land.

Farm Forestry

At first farm forestry was introduced to encourage poor farmers with small holdings of wasteland on which they took up tree planting. Seedlings fertilizers and incentives at the rate of 10 paise in second year and 14 paise in the third year are supplied from forest department. Later on, farm forestry has came to known as "group farm forestry" which means that a large number of villagers joined together and planted in one block. This is because the collective community action is more reasonable safeguard against failure. It has gain more success because of the distribution of *Patta* (land given by the Government on a 99 years lease) to the landless people to participate in such plantation activities. The superiority of group farm forestry is due to (1) Personal interest is more maintained; (2) It is less expensive; (3) It is higher in direct incentive (4) It can easily manage; and (5) Large scale contract can be given on a high price.

The actual achievement under farm forestry in the plateau districts is 96707 ha where as the appraisal target was 36,000 ha. So the actual achievement is 269 per cent. This incredible success of farm forestry is due to the high demand of timber in the timber market. The farm forestry sites are also mostly approachable which helps the buyers.

Again, plantations are mainly done on the wastelands from which they were getting negligible or no return. But from the farm forestry they used to get sizeable income.

The first farm forestry was done in Nepura village in Midnapur where now 90 per cent of the land came under group farm forestry. The community response is more in the plateau districts of the State.

Rehabilitation of Degraded Forest (RDF)

In this category rehabilitation is done both by plantation as well as through natural regeneration (coppecing). About 75 to 90 per cent of trees are fuelwood and poles and others are selected species to India. The villagers are allowed to collect 25 per cent of the produce free of cost.

The appraisal target of plateau districts was 10,000 ha which is approximately 70 per cent of the State appraisal. But the actual achievement is far ahead which is 63040 hectares.

This incredible success is due to availability of large blocks of degraded forest land. Due to maltreatment and illicit cutting fairly high percentage of forest area has attained to such a position that forests can be grown only by planting new trees and copping of sale forests.

Strip Plantations

Strip plantations are mainly raised along the roads, railways, canals, river banks etc. This plantation is fully done by social forestry wing. The villagers adjacent to the strip plantations will protect this and are allowed to collect grass, fodder, leaves and twigs and other minor forest produce from these plantations. After the sale of this forest produce the social forestry wing will recover its establishment cost and the profit will be distributed to village panchayat to establish additional strip plantations or other developmental activities.

The range of opportunity of strip plantation has to be considered in the context of large expansion of railways, roads, agricultural area etc. The railway network in the plateau districts covers about 146 km and the road system extends over about 6940 km canal and distributories also cover a considerable area.

The appraisal target of ship plantation was 10,000 ha. The actual achievement is 13323 ha. The rate is however low because of some dry and stony areas which are not available for plantation still now.

As a whole in West Bengal, the physical achievement in 72 per cent in plateau districts (Table 6.9 and Fig. 6.2).

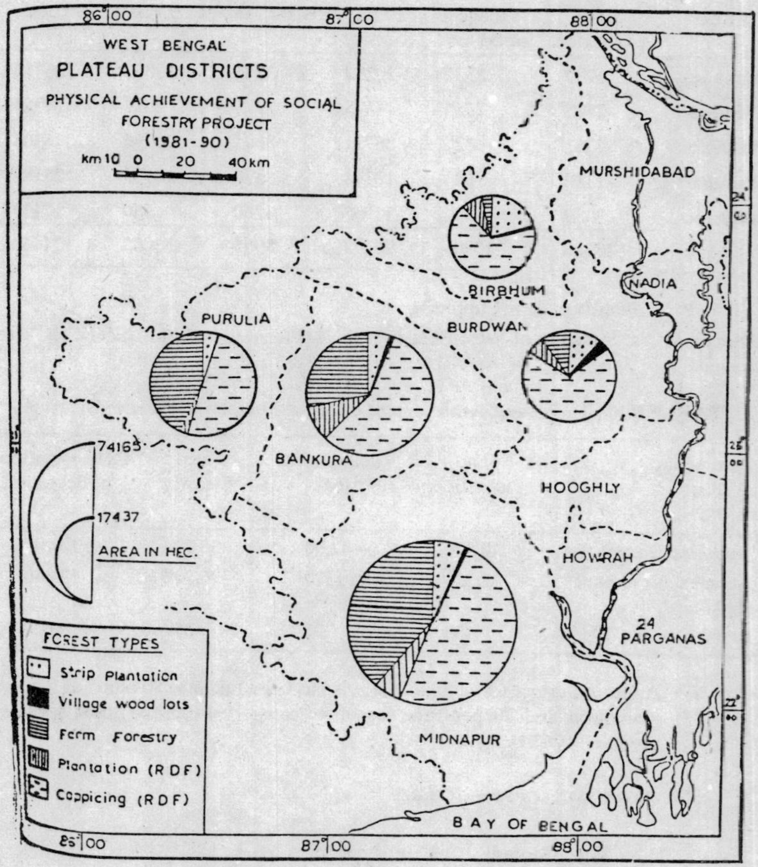

Fig. 6.2: West Bengal : Plateau Districts:Physcial Achievement and Social Forest Project, 1981-90

TABLE 6.9 : Physical Achievement of Social Forestry Project in Plateau Districts 1981-90

(*Area in ha*)

Districts	Strip plantation	Village wood lot	Farm forestry	R.D.F. (P)	R.D.F.(C)	Total
Bankura	1980	253	20282	3545	10356	36416
Burdwan	2405	262	14595	975	2618	20855
Midnapur	4361	632	36731	3645	28796	74165
Purulia	1118	60	13144	530	10735	25587
Birbhum	3459	183	11955	1260	580	17437
Total	13323	1390	96707	9955	53085	174460

P = Planting; C = Coppicing

Source : Forest Statistics of West Bengal 1990-91. Planning and Statistical Cell, West Bengal.

TABLE 6.10 : Target and Achievement of Social Forestry in Plateau Districts

Particulars	Strip Plantation	Village wood lot	Farm forestry	Rehabilitation of degraded forest
Appraisal Target (ha)	10,000	42,00	36,000	10,000
Actual achievement (ha)	13,323	1390	96,707	63,040
Achievement (%)	133	33	269	630

Note : Appraisal target for plateau districts has been taken as 50 per cent for strip plantation and 70 per cent for other category from the total appraisal target of West Bengal.

The main species selected are :

1. Eucalyptus – Eucalyptus hybrid
2. Akashmoni – *Accacia auriculiformis*
3. Subabul – *Leucaena leucocephala*
4. Minjiri – *Cassia siamea*
5. Jhaw – *Casurina equisetifolia*
6. Karanj – *Pongamina Pinnata*

The species selected for social forestry should have the following attributes. They should be:

1. Quick growing
2. Drought resistant
3. Suitable for timber, pulp etc.
4. Evergreen type
5. Suitable for firewood
6. Are not easily grazed/browsed
7. Can grow in any degraded condition of land.
8. Can be worked on short rotation
9. Preferably Coppicable
10. Leguminus

Some other species are also selected for the programme like fruit trees, viz. papaya, guava, cashewnuts in parks and gardens Amongst them, cashewnuts deserve special attention because it fetches good market.

In afforestation, eucalyptus usually gets preference. But subabul has also several advantage, some of them better than eucalyptus. Subabul is an evergreen tree originated thousands of years ago in Mexico. It can give high protein, fodder, fuel, timber and organic manure. They usually attain 2-2.5m. in height and 4-5cm. in diameter at breast height within a year. It can be grown on the wastelands. The trees become viable sources of raw materials within 3-5 years. The tree is very fast growing and capable of generating flow of multiple products. Additionally, being a leguminus species, this tree is ecologically desirable. There is scope for more extensive rise of this species in afforestation work.

It has been found from an analysis that survival rate is higher in the laterite zone 66 to 93 per cent while it is 50 to 81 per cent in Alluvial zone.

The perception for plantation is increasing very fast. Today the perception rate is 93 per cent which was only 73 per cent in 1983.

The effect of social forestry can be reveal from a field study carried on in Nepura, the first farm forestry village in West Bengal. Nepura is located in the laterite tract at Midnapur district in West Bengal.

Sadhan Ch. Pan was the first tree planter. In 1980 ha planted 3000 eucalyptus trees on one acre of *Rayat* land and 0.75 acre of *patta* land (Rayat land is superior quality land owner through inheritence or purchase while *patta* land is given by the Government on a 99-year lease). In 1985 he get Rs. 56,000 by selling 2200 trees.

TABLE 6.11: Characters of Main Plants in Social Forestry Project in the Plateau Districts

Species	Quick growing	Drought resis-tant	Suitable fortim-berpulp etc.	Evergreen type	Suitable for fire wood	Are not easily grazed/ browsed	Can grow in any degrad-ed con-dition of land	Can be worked on short rotation	Preferably coppic-able	Leguminus	Fuelwood value k Cal/ kg	Other uses
Eucalyptus	✓	✓	✓	✓	✓	✓	✓	✓	✓	×	4800	Oil
Akashmoni	✓	✓	✓	✓	✓	×	✓	✓	×	×	4800	Oil
Subabul	✓	✓	✓	✓	✓	×	✓	✓	✓	✓	4200/ 4600	Fodder Fertilizer
Karanj	✓	✓	Timber	✓	×	×	✓	✓	✓	×	4600	Oil fodder

The lands devoted for plantation are unfit for agricultural use and remained fallow for several years with that earning he purchase 3.8 acre of peddy land by Rs. 46,000 and install a shallow tubewell by Rs. 10,000. Hazra of that region planted 160 Eucalyptus trees in 1980 on 0.2 acres of rayat land and sold it at Rs. 6,000. More or less all the cash return are mainly used for down stream activities as (1) Purchase of land; (2) installation of shallow tubewell; (3) marriage of daughter; (4) construction and repair and (5) meeting up other family needs.

Arabari is another village in Midnapur where farm forestry of social forestry wing developed very much. The unproductive agricultural lands are uniquely used by afforestation and now it become a means to improve the livelihood of poor people.

Wasteland Development in Plateau Districts

Degraded land covers half of the total area in the plateau districts. These lands are normally referred as 'wasteland'. In 1985 with the establishment of National Wasteland Development Board, West Bengal Waste-land Development Corporation was formed. But the real work was started in March 1990 in a separate infrastructure under Wasteland Development Corporation limited. In West Bengal the programme was first implemented in Purulia, Midnapur and Bankura. The objectives are:

1. Plantations on the degraded forest lands for small timber, fuel wood and fodder by quick growing species on short rotation (20,000 ha).
2. Cashewnut plantation (5000 ha).
3. Bamboo plantation (5000 ha).
4. Farm forestry on 60,000 ha.

The programme will be implemented 40 per cent on Government land and 60 per cent of private lands of the *Patta* holders. The wasteland development programme will be 100 per cent financed by Central Government. National Bank for Agriculture and Rural Development will also sanction loan for this programme.

The principal aim is to develop integrated land use planning capability and to take up specific activities for conservation of ecologically fragile water-sheds, regeneration of degraded forests etc.

7

MANAGEMENT OF FOREST RESOURCES

History of Management

In ancient days, the western part of Bhagirathi was known as Radha and Vajjavumi. The eastern part of the districts of this area with its level soil, well watered by rivers, formed part of Radha. The rugged country on the west was known as Vajjavumi i.e., the country of thunderbolt. Radha (Gangaridae at Meghasthenes) was the eastern part of the territory ruled over by the Mauryan emperors and then by the emperor of the imperial Guptas of Sasanka and Harshavardhan. After 7th century, it came under the Pal Kingdom, then under the Cholas and Orissa Kings (11th century) till the middle of the 12th century and after that it passed to the Sen kings. In the 13th century, the tract passed to the Muslim rulers. But it was not so important because there were several local principalities who control the forests. The chief among them were the Malla kings of Bishnupur, Rajas of Birbhum etc. in the north and the Rajas of Panchet, Jhalda, Dhalbhum in the south. Up to 16th century Pathans hold the territories. In the Mughal period, the former one was known as Doman-Koh (Birbhum-Dumka) and the hilly jungle portion as Jharkhand. Up to the middle of the 18th century, it was under the Marathas jurisdiction. In 1960 it came under the British rule. In 1805 all the territories known as *Chakla Burdwan* was made into a new district called the *Jungle Mahals*. But for the administrative facilities the Jungle Mahals were broken up in 1833 and allotted to the present districts. The forests then were owned by the States and private individuals like Midnapur Zamindari Co., Maharaja of Burdwan, Nawab of Murshidabad, the Rajas of Vishnupur, Simlipal and Maharaja of Dharbhanga etc.

The past history of the forest management in India tells us that before and during the British rule, the rulers used to pursue the policy of encouraging agriculture at the expense of forests. The forests were open to all for indiscriminate use and ruthless exploitation. As a result, except in some remote places, the original form and structure of the forests have been modified by the successive events of history.

The alarming reports about the dismal decay of forests wealth, the Government of Bengal in their resolution No. 14,670 dated 25th July, 1938 appointed a Forest Committee to enquire into the causes and to suggest remedies to check progressive deforestation. On the recommendation of this Enquiry Committee, Bengal Private Forest Act was enacted in 1945 which was re-enacted after the partition as the West Bengal Private Forest Act in 1948. The Act has provisions for the conservation of private forests and afforestation in private waste lands. The forests were managed according to working plan approved by the regional forest officer. The private forests were controlled under the West Bengal Private Forest Act, 1948 up to the year 1954-55. During this period, the deterioration of the forest was to some extent restricted. But there was lack of co-operation within the owners. They are not willing to invest any money for long-term benefits to their forests. However, there were also some defects in the Act. In order to rectify the major defects in the Private Forest Act, the West Bengal Estate Acquisition Act was promulgated in 1953. This Act brought about tremendous changes in the land management of the State. The Zamindari system was abolished by this Act and all private forests have been vested with the Government since April 1, 1955 free from all encumbrances. Restrictions have also been imposed on the collection and removal of forest produce. The Divisional Forest Officer used to prepare felling scheme every year starting from 1955-56 onward, in consultation with the Conservator of Forests and put up the lot for sale in auction. The Act was extended to Purulia in 1964 when all forest areas retained by the intermediaries were vested to the forest department.

Status of Present Management

For centuries, individuals and communities have been able to live in harmony, but during the past few decades, the rapid growth of population and the higher expectations of the people generated throughout the world have led to an increase in the use of forests to such an extent that forests have been almost ruined. So with the growing scarcity of valuable timber and other forest products and observing bad effects upon climatic conditions, the need for scientific management and conservancy has been felt. Measures are being taken towards that end.

In the West Bengal Estate Acquisition Act (1954) forest lands were divided into reserved (5%), protected (78%), unclassed state forests (17%) etc. Steps are also taken with the object of preventing the destruction of private forests.

Reserved Forests

In the reserved forests, private rights are permanently demarcated and efforts are made to prevent damage. All rights are prohibited except which is permitted. Species are mainly reserved for timber production. With scientific management, regeneration of valuable species are carefully attended. Except in Bankura, reserved forests are found everywhere. In Midnapur it covers a limited area. In Purulia reserved forests cover only about five per cent of the total forest area.

Protected Forests

According to the *Imperial Gazetteer of India* arrangements in protected forests are less than the previous one. In this forest area, every thing is permitted except which is prohibited. Forest departments are mainly directed to the prevention of reckless felling. The protected forests cover large segments of Midnapur and Bankura District. About 78 per cent the total area is under protected forest.

Unclassified State Forest

The unclassified State Forest is managed and regulated by executive orders. The main vegetation are bamboo, palms, fruit trees, grown by the villagers which meet their ordinary requirements. In some areas good patches of forests are also found. Due to the miscellaneous species and degraded forest, unclassified state forest holds a significant position. Bankura have the highest area of unclassified state forest (Table 7.1). About 17 per cent of forest of this district belong to this type. In 1960-61 there were forest areas owned by civil authorities but after that all are vested to the government.

TABLE 7.1 : Classification of Forest Areas in Plateau Districts of West Bengal

(in hectares)

Year	Reserved Forest	Protected Forest	Unclassified State Forest	Forest owned by civil authorities	Total
Burdwan					
1960-61	331	17,847	3,200	9,516	30894
1970-71	329	21,359	6,831	–	28,519
1980-81	324	21,359	9,261	–	30944
1990-91	300	19,200	8,200	–	27600
Birbhum					
1960-61	916	3,072	7,534	2,177	13,699
1970-71	2,639	3,745	8,750	–	15,134
1980-81	2,900	5,191	7,667	–	15,760
1990-91	800	5,400	3,700	–	15,900
Bankura					
1960-61	–	1,34,747	982	4,489	1,40,218
1970-71	–	1,39,238	1,184	–	1,40,422
1980-81	–	1,24,565	14,988	–	1,39,553
1990-91	8,000	1,31,100	9,100	–	1,48,200
Midnapur					
1960-61	2	1,42,921	24,389	1,169	1,68,481
1970-71	2	1,68,764	3,316	–	1,72,082
1980-81	236	1,70,098	1,810	–	1,72,144
1990-91	800	1,16,600	53,500	–	1,70,900
Purulia					
1960-61	10,327	77,803	–	–	88,130
1970-71	10,326	72,340	4,948	–	87,614
1980-81	10,326	72,878	4,410	–	87,614
1990-91	11,200	72,900	3,500	–	87,600

Forest Management Programme during Plan Periods

After taking the possession, West Bengal Forest Department for better management, divided the forest tracts into several segments. The big districts were divided into two divisions managed under Divisional Forest Officer (DFO) as in case of Midnapur and Bankura. Small districts like Purulia, Birbhum, Burdwan are left as such and also managed by District Forest Officer (DFO). The districts/divisions have been further divided into ranges, being controlled by a Range Officer. The ranges are further sub-divided into beats managed by beat Offiicer (Table 7.2).

TABLE 7.2 : Average Range and Beat Areas in the Forest Divisions

Divisions	Forest area (in sq.km)	No. of territorial range	No. of territorial beat	Average Range area (in sq.km)	Average Beat area (in sq.km)
Bankura(N)	735	6	36	123	20
Bankura(S)	747	7	33	107	23
East Midnapur	815	8	43	102	19
West Midnapur	894	10	46	89	19
Burdwan	280	6	16	47	18
Purulia	876	9	45	97	19
Birbhum	159	5	16	32	11

Source : West Bengal Forest Statistics, 1991.

For further development, Forest Development Corporation has been established with a view to increase and to take proper care for minor forest products and speeding up afforestation programme. Under the direct management of the Government, the forest development are carried in successive Five Year Plans in the following manner.

The First and Second Plan (1951-61)

In the first plan, the focus was on afforestation and rehabilitation of the degraded forests and experiments for the plantation of economically important species. During the second Plan importance was mainly on the conservation on wild life.

The Third Plan (1961-66)

During the Third Five Year Plan, special emphasis was given to meet the long-term requirements of the country. Measures have also been taken for more economic and efficient utilisation of valuable forest products. The immediate objective was to increase the output from the existing forests through better techniques, improvement in communication and by preservative treatment. To attain self-sufficiency in industrial timbers, fuel wood and other forest products, plantation of quick growing species with high productivity rate are being carried out in large scale. With the collaboration of United Nations Special Fund (UNSF) and the Food and Agriculture Organisation (FAO), the pre-investment survey of forest resource was initiated. The training of forest officers, the executives in the State Government and forest contractors has been initiated. Such facilities are provided with the establishment of logging training centre.

The Annual Plan (1967-69)

Within this period emphasis was mainly given on plantation of quick growing species and for development of plantation techniques.

The Fourth Plan (1969-74)

The main objectives of the Fourth Plan is to increase the productivity of forests and to link it up with various forest based industries as a support to rural economy. Measures have been taken to meet the immediate and long-term agricultural and industrial requirements. Emphasis is also given on the production of raw materials for major forest based industries such as pulp, paper, newsprint, wood panel, etc.

The Fifth Plan (1974-79)

The primary objective of the Fifth Plan is to take up a dynamic programme of production forestry, through clear felling of degraded forests and creating large scale man-made forests. Other important objectives are to develop farm forestry with the help of local people. In this period, Forest Corporation has been created for further development of long-term production and production of minor forest products.

The Sixth Plan (1980-85)

The Sixth Plan brought revolution in the forest management. Importance is mainly given for conservation of existing forests, launching of country-wide afforestation and the establishment of the new concept of social forestry programme. In addition to environmental balance attention is given on conservation of natural resources, production of fuel, fodder and other domestic needs, raw materials for large and small scale industries etc. The main objective is the popularisation of appropriate management system.

Seventh Plan (1985-90)

All efforts have been directed to ensure forest protection, research and training and afforestation on largle scale through peoples participation. The main objectives and strategy in the forestry sector have been conservation of ecologically fragile ecosystems and preservation of biological diversity in terms of fauna and flora.

1. increasing substantially the vegetative cover by massive afforestation through social forestry, farm forestry and other plantation programmes.
2. meeting the basic needs of the people in respect of fuelwood fodder, minor forest produce and timber.
3. ensuring close linkages between Forestry Programmes and Welfare of the tribal and other communities traditionally depending on forests.
4. special emphasis on forestry research, education, training and extension.
5. implementing the National Wild Life Action plan for wild-life conservation.
6. creating a massive peoples movement for achieving the above objectives.

The physical management of the forest is now carried as per the prescribed silvicultural system. Mismanagement and coppicing for long years degraded the forest to such an extent that maintenance of coppice sal is not financially viable (Table 7.3). For these reasons different working circles have been formed. The main objective of it is to bring a normal age gradation to these irregular forests and to meet on sustained basis the demand of pole, raw material for pulp and paper, and firewood. For these reasons plantation under clear felling system and restocked by coppicing are carried on. Plantations of miscellaneous species will however be clear felled and artifically regenerated except in areas where natural regeneration is promising. The areas which are suitable for growing quality timbers will be used to raise timber forests by artificial regeneration (Table 7.4).

The Working Circles

To manage the forest in a systematic way four/five working circles are formed, viz., sal coppice working circle, pulpwood working circle, sal high forest working circle, timber conversion working circle, casuarina working circle.

Sal Coppice Working Circle

This working circle is formed to produce sal poles. The areas with potential but degraded sal, which by protection and cultural operations will behave like other growing areas are brought under this circle.

Pulpwood Workiing Circle

This working circle is formed to select suitable sites for quick growing species to produce raw materials for making pulp. Poor quality sal coppice forest, scrub and jhanti forest, blank and waste lands having less than 200 viable sal stool per hectare and all older plantations are included in this working circle.

Sal High Forest Working Circle

Within the sal coppice forests, some areas of well grown patches are kept reserved in this circle. This will make provision for supply of timber for local consumption and fuelwood. Alongwith it, the well grown stocks at different places will offer in future a good natural habitat for wild life.

Timber Conversion Working Circle

This working circle is created with a view to provide quality timber for various industries and also for constructional purpose. The tracts selected for this working circle consist mostly of the hilly terrains of Midnapur, Bankura and Purulia. These forests, because of their placement and physical feature will offer good refuge for wild animals and centre for tourist attracton.

TABLE 7.3 : Management of Working Circles (WC) in the Plateau Districts of West Bengal

Name	Sal Coppice W.C.	Pulpwood W.C.	Sal high forest W.C.	Timber conversion W.C.
Reasons for formation	For supplying pole, firewood	To supply industrial raw materials	For local timber prod. and natural habitat for wild life.	For production of sal and other important timbers
Method of Treatment	Simple coppice system	Afforestation and coppice regeneration	Intensive reservation and improvement in felling	Clearing degraded forest and afforestation
Constitution	The coppice sal forest	Mixed scrub blank + waste lands	Good and potentially good sal coppice areas	Hilly areas and good areas of plantation except eucalyptus

TABLE 7.4 : Distribution of Working Circles in the Plateau Districts

Name of working circles (area in ha)	East Midnapur	West Midnapur	Bankura	Purulia
Sal Coppice W.C.	–	44,176.54	69,226.65	13,799.48
Pulpwood W.C.	49,142.38	19,502.03	49,374.55	36,595.97
Sal High Forest W.C.	–	8,392.71	8,436.31	–
Timber Conversion W.C.	26,797.51	16,173.67	6,610.35	28,089.17

Source : Forest Survey of India, 1984.

Under the Sal Coppice Working Circle, maximum area is earmarked in Bankura. East Midnapur have no area under this working circle. Due to the absence of good forest in East Midnapaur Division, area under Sal high forest is also absent. West Midnapur have considerable areas under both the circles. As the timber conversion working circle covers the highest east in Purulia, the areas are not devoted to sal high forest working circle. Timber conversion working circle covers a significant area in all divisions.

Casurina Working Circle in East Midnapur covers 1,449,30 ha of land where forests are formed by the plantation of casurina trees on the Digha sea beach.

Management Problems and Proposals

Up to the past decade of this century, forest was considered to be only raw material for construction, small scale industry, wild life sanctuary etc. and there was no awareness about its conservation. Recent increase in population and rapid decrease in forest areas make the government conscious about their preservation. Several planning programme and development schemes have been chalked out for proper management of forests. But there are certain problems which hinder their progress.

Low Productivity

The main problem is low productivity. The average production is about 17m^3/ha. But it is seen that the out-turn from good coppice coupe areas are almost three times than the average i.e., 66m^3/ha. The density of stems/ha is also very low 368 stems/ha which discourage the total production. Due to the dearth of research work, the potentiality of minor forest products are not fully utilized. The proper utilization of it could help to boost up the economic condition of the people.

Lack of Investment

In earlier days, it was believed that coppice forest does not require any nourishment. But systematic coppicing helps in maintaining a good forest land. It has been found that such forests needs care, fertilizer -dressings etc. After certain period these forests require replantation.

In the plateau region, sal plants are occasionally infected with fungus in the core-wood after 15 years or so. Such trees should be removed to maintain the quality of other sal trees. Termite attack is a common problem in this area. So proper care should be taken to protect the plants.

Inadequate Staff

Though physical protection has given to the coppice forest, it is inadequate. Approximately 2,000 men are recruited as protective staff. But compared to the total forest area of about 400,000 ha it is only one person/200 ha. Even then the highly skewed boundary line of the forest areas which are often protruded in the shape of tongue in agricultural land are obviously difficult to protect with this minimum protective staff.

Moreover, the forests are scattered far and wide in small blocks. As for example, Bankura has the highest proportion of its area devoted to forest. But not more than 50 per cent of the forest areas are concentrated in big blocks say 2-3 sq. km and above. Where the forests are existing in big blocks they have an extremely logged boundary. Cultivation has been extended as extremely narrow strips within forest areas. Such irregular distribution in very small blocks are invitations for encroachment. To check such practice, steps should be taken to post 5 or more forest guards in one beat. There should be provisions for mobile petrol parties. The peripheries should be afforested with fuelwood and less valuable species. The valuable species should be kept in the centre. In such cases people generally concentrated on the peripheral section for the collection of the firewood etc.

Defective Management

Though the forest management policy has recently been improvoed, it needs further improvement. Emphasis should be given in accelerating Production Forestry Programme to meet the demands of wood-based industries, who are currently facing acute shortage of raw materials. Due to defective forest management, forest offences are at a large number.

From Table 7.5 it is clear that number of illicit felling which is most dangerous is very high particularly in good forest areas.

Illegal collection of wood often covers 10 to 20 per cent of the total revenue.

Proper management should give emphasis on research, and

techniques in adoption of better scientific techniques to foster plant growth, nutrients requirements of the plants and associated factors.

TABLE 7.5 : Number of Forest Offence during 1990-91

Division	No. of illicit felling	No. of unauthorised grazing	No. of other offences	Total
East Midnapur	1128	–	–	1128
West Midnapur	1703	13	177	1893
Bankura(N)	1299	-	11	1310
Bankura(S)	629	15	217	861
Burdwan	496	32	30	558
Purulia	847	100	10	957
Birbhum	481	69	168	718

Source : Forest Directorate, Government of West Bengal, 1991.

In the face of the recent change in society, fresh outlook is necessary. The fiscal policy should be well defined. There should be change in the stereotyped working plants and budgets.

Financial Problem

This area is facing financial crisis, thereby hindering development programmes. An analysis of the total forestry development expenditure during the plan period indicates that only one per cent of the total plan expenditure of the State is devoted to the development of forests, which is deplorably low when compared to the other sectors of her economy. The investment per hectare was negligible upto Fifth Five Year Plan. This has been marginally increased during the Sixth and Seventh Plan, when additional capital is invested on social forestry programme (Table 7.6).

Up to Seventh Plan only 1.45 per cent of the state outlay is devoted to forestry sector which is very much negligible.

The cumulative effect of the continuous low investment in the forestry sector has had its impact on the total economy. The contribution of forest to develop the rural community and pollution control are enormous. So the development of forests should be viewed not only from the commercial point of view but also from the view points of social needs and ecological requirements.

TABLE 7.6 : Five Year Plan Outlays in Forestry in West Bengal

Plan period	Total for State under dev. Plan	Forestry Sector	Percentage of forestry sector *vis-a-vis* state outlay
First Plan (1951-52–1955-56)	67.71	0.53	0.78%
Second Plan (1956-57–1960-61)	148.57	1.49	1.00%
Third Plan (1961-62–1965-66)	304.75	2.59	0.85%
Annual Plan (1966-67–1968-69)	161.46	2.49	1.54%
Fourth Plan (1969-70–1973-74)	363.46	1.96	0.54%
Fifth Plan (1974-75–1978-79)	1248.93	5.72	0.46%
Annual Plan (1979-80)	383.21	3.60	0.94%
Sixth Plan (1980-81–1984-85)	2433.27	30.18	1.24%
Seventh Plan (1985-86–1989-90)	4125.00	60.00	1.45%

Source : Forest Statistics of West Bengal, 1987.

Lack of Motivation

The tribals who are living in and around the forest areas are enjoying all sorts of privilages from the forest. Some of these privileges are no longer available to them. So the tribal people have developed a hostile attitude to the forests. In such condition, meaningful forest management can hardly be envisaged, unless the people who live near the forest and make their living out of it are benefited by appropriate management policies. There is a gap between the management staff and the local people. Hence the forests loss the support of the people. Poverty compels them to turn on the forests as the ultimate resort. As such, there should be corresponding growth of other sectors of economy to lesson the dependence of the people on forests.

8

CONSERVATION OF FORESTS

"Man is an important part of the biotic environment. His activities effect the developmental succession of forest communities" (Stabbing, 1921-26; Mobs, 1941; Misra and Puri, 1955 and Puri, 1969). It has been found from historical sources and from distribution of existing forests, that the plateau region was covered with dense forests in pre-historic and historic times. Successive waves of invasion and migration have had considerable influence on the primeaval communities of plants.

During the period of Mauryan Empire, the celebrated minister of Chandragupta, Kautilya in the *Arthashastra* mentioned the following forests viz. :

(i) Forests set apart for use of the Brahmins;
(ii) Reserved forests for the supply of all kinds of forest produces;
(iii) Elephant forests, where elephants were kept for the army and were allowed to breed there; and
(iv) Sporting forests for general public.

During the reign of Ashoka, there were mentions of some damages of the forests areas due to encroachment of villages and cultivation on the forest areas. During the Muslim period, there was considerable destruction of forests. It is also partly caused by indigenous population who fled to the forests and frequently took to shifting cultivation, a system, which is probably as destructive to forest as any other acts of man.

The Mughals were good administrators but they were not particularly interested on this resource.

During the British period, very little attention was paid to the forests in early days of their rule. But in the later period in view of poor vegetation cover, a policy of forest conservation was adopted.

Soil and Water Conservation

The degeneration of forest is gradually exposing the soils to erosional agents. With a view to check further degradation, soil conservation measures are therefore necessary. These measures aim at safeguarding all types of lands against impoverishment caused by (i) excessive soil removal, (ii) deposition of the eroded materials, (iii) exhaustion of plant nutrients through leaching, excessive cropping or over-grazing, (iv) improper cultivation, (v) burning, and (vi) accumulation of toxic salts etc. In this connection, the State Government for the first time took initiative for conservation. In 1938 a Committee known as West Bengal Forest Committee was set-up to investigate the situation and to make recommendations. The main objective was to suggest measures for the conservation of existing forests and afforestation. A Soil Conservation Circle was created with a view to afforest the wastelands to check soil erosion and flood (Table 8.1).

TABLE 8.1: Categories of Soil Erosion in Forest Area in the Plateau districts

(in sq. km)

District	Ravine erosion	Sea coastal erosion	Regular flood affected area	Erosion due to faulty quarrying	Sheet, rill and gully erosion	Total area affected
Burdwan	8.00	–	–	0.80	74.00	82.80
Purulia	5.00	–	–	–	297.00	302.00
Birbhum	1.00	–	–	1.00	10.50	12.50
Midnapur	19.50	8.00	0.40	–	1281.42	1309.32
Bankura	–	–	–	–	1255.00	1255.00

	Forest area	Percentage of forest area affected
Burdwan	277.00	29.89
Purulia	876.00	34.47
Birbhum	159.00	7.86
Midnapur	1709.00	76.61
Bankura	1482.00	84.68

From the previous Table 8.1 it has been found that the erosion has already grip quite a high percentage of area. The picture is most serious in Bankura and Midnapur where soil erosion covers most of the forest lands. To combat with this problem in this dry zone, soil conservation work however centres round special soil working techniques with particular emphasis on conservation of moisture. This must take into account the physical and structural properties of the soil as well on its nutrients. Mainly two types of land need urgent attention namely, (i) the undulating laterite soils of Midnapur, Bankura, Birbhum and Purulia Districts; (ii) the Coastal Sands in Midnapur coast. In former case, although annual rainfall is not particularly heavy but water rolling over the large extent of uncovered land gathers enough speed to etch out gullies along the slopes. The gullies continue to creep-up the slopes and ramify across the land, rendering large areas unfit for any purpose.

The second type is the sandy waste along the Midnapur coast where sand under shallow coastal waters, gets exposed to sun and wind. These are continually driven inland by the relentless breeze and engulf cultivated lands and homesteads.

Recommendations of the West Bengal Forest Committee could not, however, be given effect until the end of 1945. In that time a 20-year Plan of post-war reconstruction of forest and physical resources projected by the then Bengal Government was taken up in June 1945. But the ultimate aim of the forest department was to decrease the risk of erosion and flooding all over Bengal and to bring all waste-lands under afforestation programme so as to make the Region self supporting in respect of timber, fuel etc.

Though the afforestation and erosion survey were started before, the soil conservation programme was first launched in the First Five Year Plan. During the Second Plan, soil conservation was recognized as one of the important items of development works in view of the need for preserving the fertile top soil and improving the soil moisture of the agriculture areas. It has great role in the prevention of flood.

The Soil conservation measures envisage treatment of sheet eroded areas and gullies. In the sheet-eroded areas, digging of cattle proof trenches followed by intensive afforestation and in the gully areas, construction of extensive number of check dams and diversion channels have been advocated.

Watershed Management

Watershed management is a necessity because the engineering structure alone cannot tackle the flood and siltation problem. The flood control measures should be judicious combination of both vegetative measures and engineering structures. So the watershed area has to be tackled taking the catchment as a whole and not on piecemeal basis. Afforestation of the uppor reaches of the catchment tackle the problem at the very root. The engineering measures are effective only on the lower reaches particularly in the plains. Along with the protective role, vegetative measures also give economic returns. It helps in soil and water conservation, flood moderation and improving the water quality.

The silt studies made on the river shows that the sediment discharge of the Damodar at Panchet, Mayurakshi, Kangsabati, and the Barakar at Maithon are 10.00ha, 16.43ha, 9.65ha and 13.10ha/100 sq. km. respectively. The extent of the problem in the catchment of the Ajoy river an important river of the Ganga basin, is also very high. Out of 255,000 ha of its total catchment area, about 72,000ha or one-fourth of the catchment area is exposed to high erosion condition.

In the Ganga basin, the soil conservation scheme was first launched in the Kangsabati river valley (Table 8.2). It was initiated in 1964, and continues even to this date. A substantial percentage of area has been taken under the conservation scheme.

TABLE 8.2: Progress of Soil Conservation in the River Valleys

River	Year	Physical achievement
Kangsabati	1976–77 to 1987–88	16,506
Kaligha i	1981–82 to 1986–87	12,104
Mayurakshi	1981–82 to 1986–87	8,976

Source: Forest Statistics of West Bengal. 1981-1987.

In Kangsabati river valley, 50 catchments have been identified, of which 15 critical mini-catchments are being treated on a priority basis.

The Damodar basin is being treated since 1966-67 with the help of the Damodar Valley Corporation. Of the total area of 1,860,000 hectares in the upper valley, nearly a quarter has been put under some forest cover for decades. Because of the regged terrain and the lashing monsoon rain in this region, it is essential to maintain forest cover. In recent decades, however, a progressive deforestation continues in the upper valleys. In 1972 huge areas of waste-lands particularly in the catchments of the Darakeswar, Damodar, and Kangsabati river have been afforested and works in respect of groundwater recharging by suitable erection of check dams and gully plugging structures have been carried out. In case of the Tarafeni, Dulung and Jamuna catchments, the coppice felling cycle increased to 20 years instead of 15 years. This will help in the maintenance of better forest cover for a longer period in the most vulnerable portions of the upstreams of the sub-catchments. Some ranges of Midnapur and Purulia District have been included in the timber conversion working circle, with a view grow timber species, which will be felled on longer rotation. The tops of the hillocks need to be covered by suitable species where no normal felling will be prescribed. Cars has been taken to control the finger gullies specially in the hilly regions by putting more vegetative and engineering check dams, supplemented by contour and catch water trenching and planting of quick growing soil binding species.

The Ministry of Irrigation and Agriculture, Government of India has formed a working group to assess the magnitude of the problem in the Ganga Basin requiring watershed treatment and downstream engineering work. Depending on the flood discharge data, 18 catchments including the Rupnarayan, Ajay and Mahananda in West Bengal require urgent treatment.

The project proposes to introduce intensive soil conservation measures in the catchment areas of the Kaliaghai and Mayurakshi including ground water recharging, control of gully erosion, contour trenching and afforestations, to rehabilitate these two important river catchments in the shortest possible time.

To conserve the sandy wastes along the Midnapur coast, plantation of *Casurina Equisetifolia* are mainly practised. Sturdy two years old plants (1 mt high) are planted at a spacing 1.6 by 1.6m on the inner dunes and protected by barbed wire fencing. Two other species which have so far been found to do well are *Accacia monitiformis* (one of the wattles) and the *Anacarduim occidentale* (Cashewnut).

Though several soil conservation measures have been taken up by the government the programme has been rightly criticised as superficial, over-simplified and casual, requiring drastic revision. Soil and water conservation is a national problem but ultimately it becomes the problem of individual farmers who holds land and lives on it. They have to adopt various soil conservation practices, so as to use it for maintenance of his land for increasing production. A successful soil conservation programme is thus to be developed as people's programme.

9

SUMMARY AND CONCLUSIONS

It has thus become apparent from the study that forest resources of the plateau districts of West Bengal during the past few centuries have been subjected to continuous depletion. This has resulted in considerable decay of this wealth and the adverse effect on the environment as well as on the over all economy of the local people.

The plateau districts during the historic past was well known for its dense vegetative cover. Even during the colonial period a segment of the plateau comprising the sections of Burdwan, Bankura, Purulia and Dhanbad districts were known as *jungle mahals* presumably because of dense forest cover. Exploitation of the forest in this section began with the introduction of shifting cultivation by the tribals who are the original settlers of this region. The process received further momentum with the gradual in flow of people from the plains, initially for the purpose of mining and then to other occupations. Increasing population in the plateau section created imbalance between the demand and productive capacity of the local resources. This naturally has led to progressive decay in income level of the people resulting in unemployment and underemployment within the tribal and non-tribal people. It is quite natural that in the absence of any paying occupation, bulk of the people had to depend havily on the forest resources to sustain themselves.

Another basic reason of the rapid depletion was the colonial policy to expand their administrative influence in remote areas. A network of roads and railways were constructed through the area to strengthen the the colonial rules in these outline areas. These programme gained further momentum in the wake of tribal uprisings during the middle of the nineteenth century and the subsequent efforts of the colonial rulers to crash them. One of the measures was to clear the jungle hideouts of the national or state tribal leaders.

After independence attention has been given to check these process and various measures for regenerations of forests have been taken. But at the same time, the development programmes like the river valley projects (Damador valley, Kangsabati) and opening of new mining and industrial areas have to be implemented on areas which were formerly covered with thick forests.

This process naturally results in the further depletion of the natural vegetation cover of the region. The net result is that only 13 per cent of the total areas of the plateau is said to be covered under forest at present. But the actual area under foliage cover is said to be much less. This has naturally created the environmental problems of severe magnitude, lack of foliage cover results in much of the bare lands exposed to the direct vagaries of erosion hazards. The eroded materials were carried to the river valleys which in their process were gradually becoming shallow and shallower. The water holding capacity of the rivers are gradually diminishing and floods are become more frequent in recent years then before. The upper valley section which were not exposed to such bazards are now also experiencing limited floods. The moisture holding capacity of the river catchment is considerably less which creates water scarcity for both plant and other living bodies.

The overall economic conditions of the people either in agricultural or non-agricultural sectors, have not recorded appreciable signs of improvement. The continued rise in the price of fuel as coal, kerosence oil, have made the people more and more dependent on firewood to meet his daily requirements. So the Government policy of keeping off the forest protected or out of bounce of common people literally fails. For the absence of alternate fuels and building materials, the poor people are obliged to depend on forest, in some cases evading the law. While others are taking the measures of illegal cutting of trees in collaboration with some dishonest people of the forest department. Thus though the peripheral area of the forest are protected or little disturbed, much felling is taking place in the remote interior. So the conservation oriented forest policy literally fails due to lack of support from the common people.

Realising these difficulties, the conservation-oriented forest policy has been supplemented by social forestry since the Fourth Plan onwards. In the social forestry programme emphasis has been given to motivate the local people to regenerate trees in association with the froest for department and at the same time permitting them to use the forest for their needs. In this programme the strategy is given on the following points, (i) The production of valuable timbers from the forest department and by the villagers in their respective areas, (ii) The villagers will be motivated to generate less valuable timbers which are capable of attaining maturity within a period of 5 to 6 years. The main purpose of such programme is to meet the fuelwood requirement of the people; and (iii) The social forestry programme may also be extended to non-forest areas wherein trees can be grown in the boundary lines of agricultural fields, along the road sides or canals or rivers and may also within the farmers house. The young sapplings necessary for the purpose are provided by the State Forest Department free of cost or at nominal price. The progress attained in such programme atain no doubt time higher than any other development programme.

It has thus become apparent from the study that degeneration of forest is a comparatively easier task but regeneration is a complex and time consuming process in this depleted plateau section. But it can be successfully implemented with the jeal and efforts of the implementing authorities in association with the local people. The diffusion process requires concentrated efforts, readily available extension facilities and selection of right species suitable for different ecological niches. Side by side it is also necessary to take measures for the over all economic development of the region.

The above position of the forests in the plateau districts, the following can be granted as suggestions which could helped the planner to some extent.

In view of the circumstances of the Indian climate and general demand of the agricultural villagers, Howard (1944) came to the conclusion that the minimum land under properly managed. Forests should be 20 to 25 per cent of the total area which is far away from actual forests of plateau region: Stracey (1940) suggested that land utilization should be based on the slopes of the ground-slopes up to 15 per cent to be utilised for agriculture, 16 to 30 per cent grazing and above 30 per cent for forests. According to him about 50 to 50 per cent of the hilly regions like Purulia should be put under forests for productive purposes.

According to Krishnamurti (1952) in any region where the distance between successive contours are less than 1/13 of an inch.

the area should be under forest. According to the topographical map prepared by the Survey of India, the contour line interval is 50 ft. A 12 per cent slope corresponds to a difference in elevation of 635 ft. in a mile. Thus there will be 635/50, or nearly 13 contour lines in an inch area where the slope is 12 per cent. The areas having a slope between 12 to 20 per cent should be under grass. Moderate grazing should be allowed in such localities where erosion is not severe. When erosion is heavy, land should be dedicated to permanent forest, because in steeper slopes forests afford the best insurance against soil wash. Considering the above methods most of the hilly regions of the plain districts will have to be put under forest in this plateau region.

Intensive farming should be carried on, so that the land will be covered by crops. This coverage will protect the soil. Selection of crop and its proper rotation is also an important item. Both short-term and long-term rotations are needed. Short-term trees are needed to check immediate erosion. The species are Nepiar grass, Sabai Grass, Stylosanthis, Sishal (agave) etc. These will supply fodder and raw materials for rope or coir making.

For long-term protection, the valuable species like Sal, Sissoo, Arjun, Asan, Teak etc. are to be selected according to local environment. These will supply the necessary timber. The lands which are not favourable for the growth of valuable timber, may be used in eucalyptus plantation. With this cropping system, the balance of nature can be restored. If such measures are not immediately taken, the entire forest area of the region, is likely to be exhausted in the near future.

In addition to the above, the following will help to achieve success in forestry programme :

(i) Adequate attention must be paid to correct demarcation of forest boundaries. The unclassified state forest should be immediately converted into protected and subsequently in to Reserved Forests,

(ii) Importance should be given to proper silvicultural operations which could help to increase the production,

(iii) Forests must be managed intensively for meeting the requirements of local people. Social forestry activities which are mainly concentrated in the non-forest areas, should be directed to the forest areas as well, where the people are more dependent on forests. Social forestry activities should be taken up in the small units and in all available lands,

(iv) At present, very small portion of the total out-turn of timber, pole and firewood are consumed by the local people. So export of forest products should be controlled and made available to local people;

(v) Along with the supply of fuelwood from government, community bio-gas, coal, kerosene oil etc. should be made readily available to the people. If possible, it should be given at subsidised rates. The average consumption of coal is very low only 7.5 kg per capita per year. So there should be adequate storage of coal in accessible areas. It should be made available to the consumers at a subsidised rate. This subsidy can be contemplated against the annual loss of forest produce amounting to several crores of rupees annually, and

(vi) Though a number of schemes i.e., Drought Prone Area Programme. Integrated Rural Development Programme, social forestry programme have been taken up in the region, integrated planning is still inadequate. Steps have been taken to save the forests from forcible destruction and improving the stock and quality of the forests by the plantation progamme. But it has been found that the problem is too vast. So the co-operation is needed from all sources to achieve the desired objectives.

The region has potentiality as well as viabilities in some resources which are the products of forest. If these resources are scientifically developed and properly utilized, not only the local people will get employment but the economy of the region in general will be boosted up. The minor forest products with such viabilities are the culture of lac and tassar, extension of cashewnut cultivation, propagation of kendu leaves etc. Moreover, seeds and flowers of trees like Mahua and Sal can be effectively use for the commercial production of alchoholic beverages and edible and non-edible oils, little attention has been paid towards this end till now.

The backwardness of the region are due to long continued neglect and destruction of forest resources of the area. These can be rectified with better management technique in the lines depicted above. In the programme of such management which is necessary to involve and motivate the local people, the regeneration process.

There is no reason why the region with so many potential resources will continue to remain backward particularly when the thrust areas and the problem have been identified. The future depends on the policy-makers and the people.

BIBLIOGRAPHY

Badhwar, R.L. *et al.*, (1955), The role of minor forest products for the betterment of tribal and backward class people, *Indian Forester*, Vol. 81, No. 11, F.R.I., Dehra Dun.

Bagchi, K. and Mukherjee, K. (1978), *Diagnostic Survey of Deltaic Bengal*, Deptt. of Geography, University of Calcutta.

Bagchi, K. and Mukherjee, K. (1979), *Diagnostic Survey of Rarh Bengal*, Part-I, Deptt. of Geography, University of Calcutta.

Bagchi, K. and Mukherjee, K. (1980), *Diagnostic Survey of Rarh Bengal*, Part-II, Deptt. of Geography, University of Calcutta

Banerjee A.K. (1972), Evaporation from a young Eucalyptus hybrid plantation of West Bengal Proceedings and technical papers of symp. on Man-made forests in India, F.R.I. Dehra Dun.

— —(1973), Plantation of Accacia auriculiformis in West Bengal, *Indian Forester*, Vol. 99, No. 3, F.R.I., Dehra Dun.

— —(1979), Local needs of small timber and fuelwood in India, *Development Studies Centre Monograph*, No. 17, The Australian National University.

Banerjee, U. (1982), Constraints of social forestry in West Bengal, National Workshop on Social Forestry, Haryana.

Bhimaya, C.P. (1967), Pasture and grazing. *Proce. of Eleventh Silvi Confe.*, Vol. I, F.R.I., Dehra Dun.

Boland, D.J. (1981), Eucalyptus seeds for Indian plantation from better Australian natural seed sources, *Indian Forester*, Vol. 107, No.3, F.R.I. Dehra Dun.

Bose, A. (1966), The problem of soil erosion in the upper catchments of Kangsabati resulting siltation in the river, the measures to check it and their impact on rural economy, *Proc. of Symo.*, Directorate of Forest, Government of West Bengal.

Bose, S.C. (1968), *Geography of West Bengal*, National Book Trust, New Delhi.

Bormann, E.H. and Likens, C.E. (1970), *Pattern and Process in a Forested Ecosystem*, Springer-Verlag, New York/Heidelberg/ Berlin.

Brasnett, N.V. (1953), *Planned management of forests*, Ruskin House Museum, London.

Chatterjee, A.B. (1970), West Bengal, A Pub. of Geographical Institution, Presidency College, Calcutta.

Chakraborty, Kalyan 1980, Role of social forestry in tassar culture, An experiment in Birbhum District, West Bengal (unpublished).

Champion, H.S. and Seth, S.K. (1980), Role of social forestry in tassar culture. An experiment in Birbhum District, West Bengal (unpublished).

Champion, H.S. and Seth, S.K. (1968), *General Silviculture of India*, F.R.I., Dehra Dun.

Chatterjee, C.D. (1964), *Forestry in Ancient India, West Bengal Forests Centenary Commemoration Vol.*, Directorate of Forests, Govt. of West Bengal, Calcutta.

Chaturedi, M.D. (1951), The role of vegetal soil cover in flood control, *Indian Forester*, Vol. 77, No.3, Dehra Dun.

Chaudhury, K.N. (1957), Afforestation techniques for the lateritic zone, *West Bengal Forest Bulletin*, No.5, Directorate of Forests, Calcutta.

Collect, M.E.W. (1970), External costs arising from the effect of forests upon stream flow in Britain, *Forestry*, Vol. 43, U.K.

Dabral, B.G. (1967), Climatological studies in Indian forests, *Proc., of 11th Silvi, confc.* Vol. II, F.R.I., Dehra Dun.

Das, H.P. (1954), *The Forests of Assam—A Study in Economic Geography* (Thesis).

Desai, Vasant (1984), *Issues in Agriculture and Forestry*, Himalaya Publishing House, Bombay.

Ganguly, J.K. (1977), Preliminary studies on the adaptability of Eucalyptus under edhaphic and climatic variables *Proc. of Symp. 11th Silvi Conference*, Vol. II, F.R.I., Dehru Dun.

Ghildyal P.B. (1966), Soil structure problems in laterites of Midnapur district of West Bengal, *Proc. of Symposium*, Directorate of Forests, Calcutta.

Ghosh R.C. (1977), *Handbook on Afforestation Techniques*, Forest Research Institute, Dehru Dun.

– – (1977), Study of moisture relation in sal forests, *Proce. of 11th Silvi Conf.* F.R.I., Dehra Dun.

– – (1981), Management of tropical forest in India and its socio-environmental impact, Proceeding Divn. I, XVIII, *IUERO World Congress*, Intenational Union of Forest Research Organisation, Japan.

Ghosh, R.C., Subba Rao, B.K. (1979), Forests and floods, *Indian Forester*, No. 4, Vol. IV, F.R.I., Dehra Dun.

Government of Gujarat (1978), *Social Forestry Project*, For World Bank Assistance, Directorate of Forest, Gujarat.

Government of India (1954), Forestry in West Bengal, Published on the occasion of the *Fourth World Forestry Congress*, F.R.I., Dehra Dun, December.

– – (1983), "Midnapur", Report of the Forest Survey of India, Eastern Zone, Ministry of Agriculture, Calcutta.

– – (1983), "Bankura", Report of the Forest Survey of India, Eastern Zone, Ministry of Agriculture, Calcutta.

– – (1983), "Purulia", Report of the Forest Survey of India, Eastern Region, Ministry of Agriculture, Calcutta.

Government of West Bengal (1939), Report of the W.B. forest committee, Forest and Excise Department, Calcutta.

– – (1980), Constraints to forest development in West Bengal, Social Forestry Project, Forest Directorate, Calcutta.

– – (1981), Decuments of the World Bank, West Bengal Social Forestry Project, Staff appraisal report, Report No. 3434, Directorate of Forests, Calcutta.

– – (1981), Report on Social Forestry Project in West Bengal, Directorate of West Bengal, Calcutta.

– – (1981), The basic role of shellac export promotion council, Lac Development Corpn., Calcutta.

– – (1982), First working plan for Midnapur forest division, 1972-73 to 1981-82. District Forest Officer, Midnapur.

– – (1982), Forestry in West Bengal – An appraisal, (1977-82) Directorate of Forests, Calcutta.

Griffith, A.L. (1951), The necessity of afforestation in the Dry areas, *Proc. of the 7th Silvi. Conf.*, Forest Research Intitute, Dehra Dun.

Hane Jurgen, (1977), The contribution of forestry to regional development in the Sahel. *Applied Science and Development*, Vol. 10, G.D.R.

Hunter, W.W. (1877), *A Statistical Account of Bengal*, London; Reprint from Concept Publishing Company, 1977.

Imperial Gazetteer of India (1909), *Provincial series, Bengal*, Vol. I, USHA-1984

Ishaque, H.S.M. (1947), Agricultural Statistics by plot to plot enumeration in Bengal 1944 and 1945, Deptt. of Agriculture, Forests and Fisheries, Govt. of West Bengal

Jones, Gareth (1979), *Topics in Applied Geography: Vegetation Productivity*, Longman Group Limited, London.

Kuma Prem (1978), Environmental planning and forestry—The Canadidan experience, *Indian Forester*, Vol. 104, No. 5, Dehra Dun.

Majumder, S.C. (1942), *Rivers of Bengal Delta*, University of Calcutta, Calcutta.

Misra, R. and Puri, G.S. (1954), *Indian Manual of Plant Ecology*, Oxford Book and Stationary, India.

Misra, S.N., (1966), Introduction of Eucalyptus in the laterite tract and its impact on the economy, *Proc. of Symp., Div. of Forests*, Govt. of West Bengal.

Mocullah, Patrick, (1975), *Science in Geography, Data use and Interpretation*, Oxford University Press.

Mondal, S.S. (1977), Soil conservation and afforestation techniques for donuded and barren tracts, of South West Bengal, *Proc. of 11th Silvi, Conf.*, F.R.I., Dehra Dun.

Mukherjee, Anita (1981), Preliminary studies on social forestry activities in Purulia district, *Indian Journal of Landscape System*, Vol. 4, No. 1, and 2, Calcutta.

– (1982) *Cashewbut Cultivation in Midnapur District*.

– (1983), Impact of social forestry on local community development in West Bengal, *Geographical Review of India*, Golden Jubilee Vol., Calcutta.

Mukherjee, R.K. (1938), *Changing Face of Bengal*, University of Calcutta, Calcutta.

Mukhopadhyay, S.C. (1980), *Geomorphology of the Subarnarekha Basin, The Chota Nagpur Plateau* (Eastern India), The University of Burdwan, Burdwan.

Nashid, M.A. (1958), Farm forests and their contribution to the agricultural economy, *Proc., of Farm Forestry Symp.*, I.C.A.R., New Delhi.

Nath, B. and Kamath, P.B. (1977), Natural regeneration of sal

(*Shorea robusta*) in M.P. *Proce. of 11th Silvi Conference*, F.R.I., Dehra Dun.

Nisbet, John (1893), *Climatic and National Economic Influence of Forests*, London.

Palit, S.B. (1972), Economics of protection in the laterite forests of South Bengal, *West Bengal Forests*, Vol. 2, Directorate of Forests, Calcutta.

Paranjape, N.R. et al, (1981), Social forestry: A programme for tree pastures, *Economic Times*, 6th June, Daily magazine, Calcutta.

Pradhan, I.P. and Dayal Rajbans (1981), Farm forestry in agricultural economy, *Indian Forester*, Vol. 107, No. 10, F.R.I., Dehra Dun.

— — (1960), *Indian Forest Ecology*, Vol. I, Oxford Book and Stationary Co., Calcutta, Delhi.

Puri, G.S. (1960), *Indian Forest Ecology*, Vol. II, Oxford Book and Stationary Co., Calcutta, Delhi.

Ramzov, N.P. and Pogrebnyak, P.S. (1969), *Israel Programme for Scientific Translation*, Jerusalem.

Rao, H.S. (1967), Minor products, *Proc. of 11th Silvi, Conf.*, F.R.I. Dehra Dun.

Rao, Sitaram (1979), *Introduciton to Social Foestry*, Oxford and IBH Publishing Co., U.K.

Richard, St. Babre Baker (1948), *The Story of the Forests of the World*, Green Glory, London.

Roy, M.P. (1971), Plantation of casurina equisetifolia in the Midnapur district, West Bengal, *Indian Forester*, Vol. 97, No. 7, F.R.I. Dehra Dun.

Roy, P.K. (1966), The impact of forestry and soil conservation works on the rural economy of laterite region of the western districts of West Bengal, *Proc. of Symp., Directorate of Forests*, West Bengal.

Roy, Mukul, P., (1971), A Preliminary study of run-off and soil wash in run off plots at Jhargram, West Bengal, *Indian Forester*, Vol. 97, No. 1, F.R.I., Dehra Dun.

Roychowdhury, K.C. (1966), Afforestation and soil work technique in the derelict forest and waste lands in the lateritic zone southern circle, *Proc. of Symp.,* Directoroate of Forest, West Bengal.

Sahu, N.C. and Das, B. (1979), An appraisal of the forest policy in Orissa, *Indian Forester*, Vol. 105, No. 1, F.R.I., Dehra Dun.

Sanyal, Pranabesh (1972), A brief review on the formation of laterites and lateritic soils with a notes on its afforestation, *West Bengal Forests*, Vol. 2-3, Directorage of Forests, West Bengal.

Sarin, Sharad (1981), Management of minor forest produce, Perspective and alternative framework for research and analysis, *Indian Forester,* Vol. 107, No.7, F.R.I., Dehra Dun.

Sharma, B.K. and Verma, V.P.S. (1978), Studies on the production and collection of sal seeds, *Indian Forester*, Vol. 104, No. 6, F.R.I., Dehra Dun.

Simmons, I.G. (1931), *The Ecology of Natural Resources*, Edward Arnold London (Publishers) Ltd.

Sing, N. (1955), Role of forests in flood control, *Indian Forester*, Vol. 81, No. 4, F.R.I., Dehra Dun.

Sinha, J.N. (n.d.), Serve the tribal to save the forest, *Indian Forester*, Vol. 105 No. 4, F.R.I., Dehra Dun

Srivastav, B.P. and Pant, M.M. (1979), Social forestry on a cost benefit analysis framework, *Indian Forester*, Vol. No. 1, F.R.I., Dehra Dun.

Steinlin, Hansjurg (1979), The role of forestry in Rural Development, Applied science and Deve., A Biannual collection of Recent German *Contributions Concerning Development Through Applied Seciences*, Vol. 13, Tubingen, G.D.R.

Tewari, R.N. Mascarenhas, O.A. (1980), Community forestry management for rural development, Select papers and proceedings of the Second Workshop, Xavier Institute, Dehra Dun.

Ureshi, I.M. (1977), Some growth statistics of Eucalyptus hybrid (Mysore Gum, *Syn. E. teretocornis*), *Proc. of 11th Silvi. Conf.*, F.R.I. Dehra Dun.

Wadia, D.N. (1965), *Gazetter of India*, Vol. I. Ministry of Information and Broadcasting.

Wathanda, G.S. (1958), The role of farm forestry in our national development, *Proce. of farm forestry symp.*, ICAR, New Delhi.

Index

[illegible] Tables